52

PENGUIN PASSNO[...]

GCSE Biology

Trevor Day was educated at Nottingham University, where he obtained a B.Sc. in zoology. He took his M.Sc. in marine biology at the University of North Wales, Bangor. This was followed by a period of research funded by the United Nations Development Programme in Egypt. He has also taught biology at a number of schools in this country.

Paul Haddlesey was educated at Bedford College, University of London, and received his postgraduate Certificate of Education from the Institute of Education at London University. He now teaches in a comprehensive school in north London.

PENGUIN PASSNOTES

Biology

REVISED EDITION FOR GCSE

TREVOR DAY AND PAUL HADDLESEY
ADVISORY EDITOR: STEPHEN COOTE, M.A., PH.D.

PENGUIN BOOKS

Penguin Books Ltd, 27 Wrights Lane, London W8 5TZ (Publishing and Editorial)
and Harmondsworth, Middlesex, England (Distribution and Warehouse)
Viking Penguin Inc., 40 West 23rd Street, New York, New York 10010, USA
Penguin Books Australia Ltd, Ringwood, Victoria, Australia
Penguin Books Canada Ltd, 2801 John Street, Markham, Ontario, Canada L3R 1B4
Penguin Books (NZ) Ltd, 182–190 Wairau Road, Auckland 10, New Zealand

First published 1985
Reprinted 1986, 1987
Revised edition 1988

Made and printed in Great Britain by
Richard Clay Ltd, Bungay, Suffolk
Filmset in Monophoto Times Roman

Contents

Acknowledgements

The authors would like to thank two colleagues, Susan Jennings and Modris Bite, for helpful discussions during the preparation of this book. These colleagues and a student, Merryn Collard, supplied a number of the original drawings. Mr Geoffrey Goats of Rothamsted Experimental Station kindly provided background information on agriculture for Chapter 14.

1. How to Use This Book

These Passnotes provide a step-by-step guide on how to study, revise and tackle examination questions in GCSE Biology. Special attention is given to points the examiner is looking for, and common mistakes are pointed out. These Passnotes are not a textbook, but they do contain the basic information needed to gain a good pass. They are best used alongside your class notes when you are revising. With proper use they should enable you to enter the examination with confidence and achieve success.

GCSE Biology examinations test a student's ability to:

1. Understand essential biological principles.
2. Make and record observations accurately.
3. Design and carry out simple practical investigations.
4. Recognize relationships between various areas of biology.
5. Demonstrate biological knowledge using appropriate terminology.
6. Interpret data in the form of tables, charts, diagrams, graphs and photographs.
7. Apply elementary mathematics, chemistry and physics to biology.

First you need to know which syllabus you are doing and what will be required of you in the examination. The schemes of assessment vary from board to board, although all of them offer some practical assessment. This practical assessment will be carried out by your teachers during the course.

The various schemes of assessment are described below.

Northern Examining Association (NEA)

Paper 1 is of 2 hours duration and is taken by all candidates. It accounts for 35 per cent of the total marks.

Paper 2 is of $1\frac{1}{2}$ hours duration and accounts for 35 per cent of the total marks. There are two levels of Paper 2; candidates expected to get grades G–C take level P, candidates expected to get grades E–A take level Q.

Practical Assessment accounts for the remaining 30 per cent of the marks.

London and East Anglian Group (LEAG)

Paper 1 is of 2 hours duration and is taken by all candidates. It accounts for 50 per cent of the total marks.

Paper 2 is for candidates expected to get grades G–C.

Paper 3 is for candidates expected to get grades E–A.

Both Papers 2 and 3 are of 1½ hours duration and account for 30 per cent of the total marks. Candidates take *either* Paper 2 *or* Paper 3, not both.

Practical Assessment accounts for 20 per cent of the marks.

Southern Examining Group (SEG)

Candidates expected to get grades G–C follow a different examination pattern to those candidates expected to get grades G–A.

Grades G–C: one written paper of 2 hours duration which accounts for 80 per cent of the marks. Practical Assessment accounts for the remaining 20 per cent of the marks.

Grades G–A: Paper 1 is of 2 hours duration and accounts for 50 per cent of the marks. Paper 2 is the Practical Assessment which accounts for 20 per cent of the marks. Paper 3 is of 1½ hours duration and accounts for 30 per cent of the marks.

Midland Examining Group (MEG)

Paper 1 consists of 50 multiple-choice questions, is of 40 minutes duration and accounts for 30 per cent of the marks.

Paper 2 lasts for 1 hour 20 minutes and accounts for 50 per cent of the marks.

Practical Assessment accounts for 20 per cent of the marks.

Paper 3 is available only to those candidates expected to get grades A or B. It is of 1 hour 15 minutes duration.

Welsh Joint Education Committee (WJEC)

Candidates take one written paper which accounts for 80 per cent of the marks. Those expected to get grades G–C take Paper 1 which is of 2 hours duration. Those expected to get grades D–A take Paper 2 which is of 2½ hours duration.

Practical Assessment accounts for 20 per cent of the marks.

You also need to know which topics are required by your syllabus. The syllabuses vary quite a lot in content so it is worthwhile getting a copy of your syllabus so you know what to revise.

In your revision, you would be unwise to leave out major sections of the syllabus. A proper understanding of biological principles depends on knowing how the different processes in an organism fit together to make a living, working unit. Miss out any process and you do not have the overall picture. For example, how can you understand nutrition properly if you do not know about transport? After all, once food enters the body it must be transported to where it is needed.

You can use this book to refer to any points of difficulty during your revision. Perhaps better, the Passnotes can be used alongside your class notes throughout your revision.

Chapters 3 to 15 contain the course contents and each chapter has the following layout:

1. Introduction
2. The main body of the text, containing numbered topics and including definitions.
3. Near the end of each chapter is a list of definitions and key words.
4. At the end of each chapter are a number of relevant questions.

The list of definitions and key words allows you to test yourself on the factual contents of the chapter. The questions test your understanding and give you the opportunity to plan examination answers. It is worthwhile getting the Penguin *Passnotes Book of Revision Tests* which will allow you to try out a wider variety of questions than there is room for in this book.

The final chapter gives advice and encouragement on what to do just before, and during, your examination.

First of all however, you should read Chapter 2. This gives a comprehensive guide on how to tackle the different types of examination question you will come across. Learning and revision methods are also included and suggestions are made on how to plan and organize your study.

2. *Working Towards the Exams*

Having worked out which topics you need to know for your syllabus, and what types of question you will be asked, you should then *plan* your study and revision. To do this, follow through the suggested sequence below, point by point.

1. First of all make sure your school or college notes are complete.
2. Tidy up your files or notebooks and make sure they include the following:
(a) clear underlined topic titles and sub-titles
(b) underlined 'key words'
(c) clear diagrams with titles
(d) a contents list, referring to numbered pages in your notes.

If you have any notes missing, or have any points of difficulty, see your teacher about this.

3. Several months before the exam, plan how much revision you need to do each week to finish revising your complete course *before* the exam. Break your coursework down into major sections and work out how many sections you need to revise each week or month in order to finish on time.
4. The use of compulsory short-answer and multiple-choice questions means that exam questions can be set covering the whole syllabus. You would do well to have a thorough knowledge of all parts of the syllabus. However, if you have particular trouble understanding certain isolated topics, e.g. Genetics and Evolution (Chapter 13), you might be well advised to 'cut your losses' on some of these topics and concentrate on the rest of the syllabus. Again, you should see your teacher about this. Most of Chapters 3 to 12 is 'core' information which is necessary for a proper understanding of biological processes and will be required for you to give full and complete answers in Biology exams.

2.1. ***Revision Technique***

1. Revise the course topic by topic. You can use the section headings

and sub-headings in this book to help you. Reduce your notes or chapters of this book into section and sub-section headings. Write these down on a separate sheet of paper. For example (see page 47):

Topic: Nutrition
Sub-topics: Humans – digestion
absorption
assimilation
Plants – photosynthesis
water and mineral salts

Read over each section and sub-section several times. Make a list of the key facts. Check your understanding of all the biological terms and definitions. At this stage it may be very helpful for you to construct a pattern diagram for each topic (see page 6).

You should be able to reach the stage where you can reproduce these sections and their contents *without* prompting from the book or your notes. The sub-headings and key words you remember should trigger off a mass of related facts in your memory.

2. Having reached the stage where you can recall topics by remembering section and sub-section headings and key words, do not let your efforts go to waste. Unless you *review* your notes regularly, you will rapidly forget them. Reviewing means quickly re-reading your notes and headings, and this should be done at one- or two-week intervals. This is very important. It is only by regular reviewing that your organized knowledge will be retained in your long-term memory.

3. It is sometimes difficult to remember the list of key facts which relate to a particular topic; for example, the functions of blood (page 120) or the functions of the liver (page 82). There are certain techniques to help you do this:

(a) One way is to use a **mnemonic**. These are words, sentences or phrases from everyday language which are simple to remember; these are linked to technical words which are difficult to recall. For example, the different regions of the vertebral column and the order in which they occur (see page 193), can be remembered using the following mnemonic:

Several	**Ce**rvical
Thirsty	**Th**oracic
Lumberjacks	**Lumbar**
were	
Stacking	**Sac**ral
Coal	**C**audal

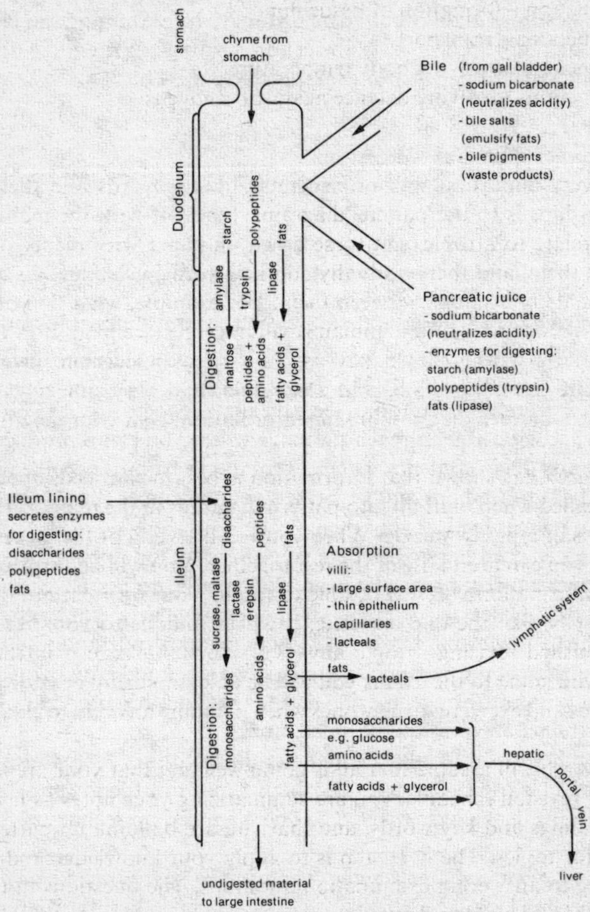

Fig. 2.1. *Pattern diagram for the human small intestine*

Wherever possible, make up your own mnemonic. Often ruder ones are easier to remember. Two more examples are found on pages 19 and 290.

(b) Another way to remember a long list is to group related items together and remember how many there are in each group. For example, the functions of blood (page 120) consist of 11 items which can be grouped into 4 categories:

1 function – formation of tissue fluid
5 functions – transport
2 functions involving body temperature
3 functions involving defence against pathogens

(c) A very important way of remembering key words and their inter-relationships is to use pattern diagrams. Start off with the main points which relate to a topic, put these down on paper with plenty of space around them, and then gradually add key words, using arrows to show functional connections between them. For example, what do you know about the functions of the human small intestine (see page 75)? You can start off by writing down key words such as **duodenum**, **ileum**, **bile**, **pancreatic juice**, **digestion**, and **absorption**, and very quickly you can build up a pattern diagram linking information from your memory. For example see Fig. 2.1.

Research has shown that information is best memorized, understood and recalled when built up into pattern diagrams in the mind, rather like the parts of a jigsaw puzzle. When you recall several of the pieces in the puzzle, you can gradually fit the rest together. You will be surprised how much information you will be able to recall. Your own pattern diagram may not be like the one opposite; it will be unique to you. When you have finished revising a topic always try to summarize it in this way. When you come to the exams you will be able to remember your pattern diagrams and even create new ones when planning answers to essay questions.

4. Now, we will assume that all is going well and that your are keeping to your revision schedule, you are summarizing your notes as headings, sub-headings and key words, and that you are building up pattern diagrams for topics. The next step is to apply your knowledge and understanding to answering examination questions. The questions in the text and at the end of each chapter are there to enable you to do this. To start with you may require prompting from your notes or this book to help you find the answers. Often, you will notice, the same question requires knowledge from different parts of the syllabus. You should reach the stage when you can answer questions without referring to your notes or to the book, and in the time allowed in the examination. Get your answers marked if you can; if not, use your notes and this book to check the answers.

The next section in this chapter looks at the different types of exam question you will be set.

2.2. *Types of Exam Question*

Short-answer questions

These make up the first part or the whole of certain examination papers. You are asked to give one-word, one-sentence or short-paragraph answers in spaces provided on the question paper. You can tell how much you need to write by the size of the space provided, and by the mark allocation. Here are two examples. Suitable answers are given.

Q. (*a*) *Why do trees usually die when a complete ring of bark is removed from round the trunk?*
The phloem is removed and this prevents organic substances such as sugars passing down the trunk to the roots. The roots eventually starve to death and cause death of the tree.

(*b*) *Give one way in which this damage could happen naturally.*
By deers eating the bark.

Q. (*a*) *Name a hormone in the body of a mammal.* Insulin

(*b*) *Name the organ where this hormone is produced.* Pancreas

(*c*) *Describe* one *effect of this hormone in the body.*
It stimulates the uptake of glucose by cells in the body and causes a lowering in blood glucose level.

In other examples, you may be given a photograph, diagram, table or graph about which you have to answer several short questions. Keep your answers simple, brief, and to the point. Be guided by the size of space provided and by the mark allocation. In general, one mark is given for each fact. A definition or brief explanation may be worth two or more marks.

Multiple-choice questions

Multiple-choice questions, sometimes called fixed response or objective questions, make up all or part of certain examination papers.

Normally, you are asked to choose the correct answer from a choice of four or five which are given. For example:

Q. $C_{12}H_{22}O_{11}$ *is the chemical formula for*
A. an amino acid

B. a protein
C. a fat
D. a carbohydrate
E. a nucleic acid.

If you do not know the correct answer right away, you can at least eliminate those which you know to be incorrect. For example, A, B and E can be discounted, because they all contain nitrogen. That leaves you with C and D, and if you blind guess, you have a 50 per cent chance of being right. You may, however, by this time, have thought out the correct answer, which is D (see page 27).

Students often find they do poorly on multiple-choice questions, although they expect to do well. Two common faults are misreading the question, or overlooking the best answer. Take great care with multiple-choice questions and double-check your answers. Sometimes the questions are rather lengthy and need to be read very carefully.

Usually multiple-choice questions ask for *one* correct answer. If you indicate two correct answers, you will automatically be marked wrong. If you do not know the answer to a question, you should choose an answer anyway. That way you at least have a chance of getting the question right and being awarded a mark.

Longer-answer questions

These make up the second part or the whole of an examination paper. There are two types of longer-answer question: the traditional essay-type question and the increasingly popular structured question.

Structured questions are made up of several parts, each of which usually requires a one- or two-paragraph answer. The length of the answer will be governed by the marks allocated to it. For example:

Q. Describe the chemical composition and biological functions of
(*a*) *carbohydrates*
(*b*) *fats*
(*c*) *proteins*
Give an account of the digestion of proteins in a mammal.
(*Details of dentition, tooth structure and digestive organs are* not *required.*)

In longer-answer questions it is important to plan your answer before writing it out fully. There are a number of reasons for this. In marking your answer, the examiner is looking for a number of key facts, clearly explained and preferably in a logical order. Planning improves your ability to come up with a really clear and complete answer, free of irrelevant information.

The best way to answer a structured question is to look for the key words in the question, underline them, and plan your answer around these. For example:

The key words relating to parts (a), (b) and (c) of the above question are **chemical composition** and **biological functions**.

The answer plan for (a) is:

chemical composition ← carbohydrates	→ **biological function**
– contain C, H, O	– broken down to release energy (monosaccharides)
– 6-C sugar units	– energy stores (starch, glycogen)
– monosaccharides ↓ polysaccharides	– in plant cell wall (cellulose)

Which can then be written out as:
(a) Carbohydrates are made up of the chemical elements carbon, hydrogen and oxygen combined to form 6-carbon sugar units. Monosaccharides, e.g. glucose, are made up of one of these sugar units, while polysaccharides are made up of many. Monosaccharides are broken down in respiration to release energy. Polysaccharides such as starch (in plants) and glycogen (in animals) are used as energy stores. The polysaccharide cellulose is the structural material in the cell wall of plants.

Now try answering parts (b) and (c). The relevant information is given on pages 29 and 30. The key words in the second part of the question are **digestion**, **proteins** and **mammal**. Construct an answer plan using the information on page 79.

In a traditional essay-type question *you* will have to provide most of the key words and will have to construct a framework around these. For example:

Q. Describe the structure and functions of mammalian skin.

A diagram of mammalian skin in section is a good starting-off point (see Fig. 8.6, page 157).

The key words in the question are clearly **structure** and **functions**. Using these as your starting point, you should be able to list other key words under these headings.

Structure	Functions
Two layers:	Five main functions:
1. Epidermis	1. Protection
2. Dermis	2. Temperature regulation
	3. Sensitivity
	4. Excretion
	5. Vitamin D production

Now, each of these key words will lead on to others, and you will be building up your essay. See pages 157–8 and build up your essay plan using as a framework the key words above. You may be able to build up a pattern diagram showing structural and functional relationships (see page 6).

By providing a plan, you are not only ordering your thoughts and planning your essay logically, but you are helping yourself to remember the facts you know. Once you have written a fact down on paper, you can concentrate on remembering other facts. Also, without your outline to refer to, key points could easily be left out of your answer.

Part of your revision should consist of preparing essay plans or pattern diagrams for questions likely to come up in the exam. Get hold of past examination papers so that you can map out specimen answers.

Questions on experiments

A description of an experiment should be written up in a logical order under appropriate headings: 'Aim of the Experiment', 'Method', 'Results', 'Conclusion'. See, for example, the experiment described on pages 53–4.

For the 'Aim', do not write 'Demonstration of . . .' or 'Experiment to prove . . .', since this implies you knew the result before you did the experiment. Instead, follow proper scientific method and write 'Experiment to investigate . . .' or 'Experiment to examine . . .'.

In the 'Method' section, where appropriate, use a diagram of the apparatus to convey clearly as much information as possible. Your account can now follow: 'The apparatus was set up as shown in the above diagram.' In your written account do not duplicate information already given in the diagram. Concentrate on any special features of the apparatus and experimental procedure, such as how often the results were recorded and how long the experiment was run for. Include in your account or diagram, where appropriate, quantities used.

Write in the past tense and use the impersonal 'was recorded' not 'I recorded'. Do not forget to emphasize which was the **test** and which the **control** part of the experiment.

In the 'Results' section a plain statement of what happened in both **test** and **control** is required, without any discussion. If the rate of a process, or an amount, is being measured, use appropriate units.

The 'Conclusion' is the answer to the aim of the experiment. This is a plain statement, not an essay. In the question:

Q. Describe in detail an experiment which would demonstrate that light is needed for photosynthesis.

the aim of the experiment would be 'to investigate the necessity for light in photosynthesis', and the experiment's conclusion is likely to have been 'light is necessary for photosynthesis'. You may have some comments or criticisms to make about the experiment. A brief mention may gain you some credit.

Diagrams

Drawing biological diagrams is a skill you should master. With patience, care and practice anyone can draw clear and accurate diagrams. They are a valuable source of marks in examinations.

Diagrams can be used in longer-answer questions whenever they make your answer clearer and save time. Alternatively, diagrams may be specifically asked for in questions, for example, 'Give an illustrated account of . . .' or 'With the help of diagrams describe . . .'.

In longer-answer questions diagrams can be used to complement written text. In the text you can freely refer to your diagram, but do not simply repeat information shown in the drawing. Remember, diagrams are supposed to make your answer clearer and save time.

Diagrams are simplified, explanatory drawings. They are not the same as the 'true-to-life' drawings done in practical classes or practical examinations (see page 17).

When drawing diagrams, take note of the following points:

1. Draw in pencil (that way you can easily rub out mistakes).
2. Your diagram should be large (at least twice the size of those shown in this book).
3. The diagram should be made up of distinct, continuous lines (do not sketch).
4. The diagram should be fully labelled.
5. Label in ink.
6. A straight line should run from the label to the structure to which it refers (do not use curved or branched lines).

7. Label lines should not cross.
8. The diagram should have a specific title.

We can apply these points in answering an examination question:

Q. Make a labelled diagram to show the arrangement of structures in the leaf of a flowering plant as seen in vertical section. Indicate on your diagram the distribution of chloroplasts.

Fig. 5.2 on page 50 is a suitable answer to this question. The upper diagram is a general plan showing the distribution of tissues in the leaf. The lower diagram is an expanded detail of part of the leaf. This shows the shape, sizes and arrangement of cells in the different tissues. Drawing many identical structures in a diagram is a waste of time. It is best to draw carefully a few of the structures and indicate that many more are present. The question asks you to indicate the distribution of chloroplasts, so obviously some detail of cell contents is required. In the example shown, only one cell from each tissue type is drawn in detail.

In drawing, use a sharp pencil – a 1H or HB pencil is best. First, map out the general shape lightly in pencil. Divide up the diagram into its major parts, making sure the proportions are correct. Fill in any detail last of all. There is nothing worse than starting at one end of your diagram, drawing in great detail, and getting to the other end and finding it will not fit. Always begin with a general plan, getting the proportions of the major parts right, and then fill in the detail last.

It is well worth copying out all the major diagrams from your class notes or from this book. Then try drawing them *without* referring to the original. Be critical when assessing your diagrams. Are they clear, accurate, properly labelled and titled?

In general, you should avoid colouring or shading. This takes time and may make your diagram less clear. In diagrams of the mammalian circulatory system, blood vessels which carry oxygenated blood can be shown by a red line, and those carrying deoxygenated blood by a blue line. If you do this, you should provide an explanatory key.

Always give a title to your drawing, and be as specific as possible. For example:

Q. Make a labelled diagram to show the structure of a mammalian neurone.

There are three main types of neurone (see page 165) and whichever one you draw, you should give its appropriate name. If you miss this out you may be penalized heavily.

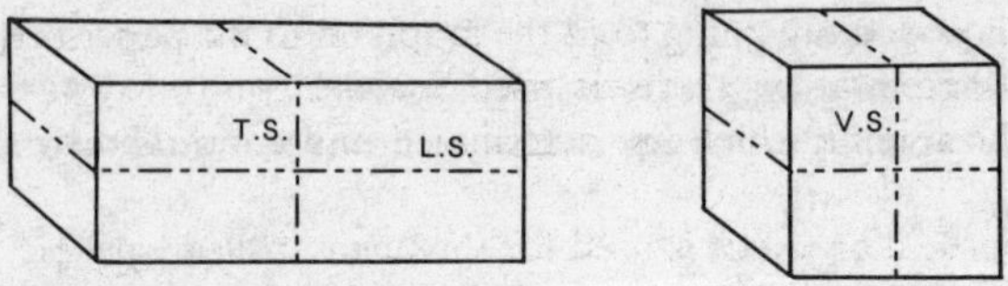

Fig. 2.2. *Orientation of sections*
T.S. = transverse section
L.S. = longitudinal section
V.S. = vertical section

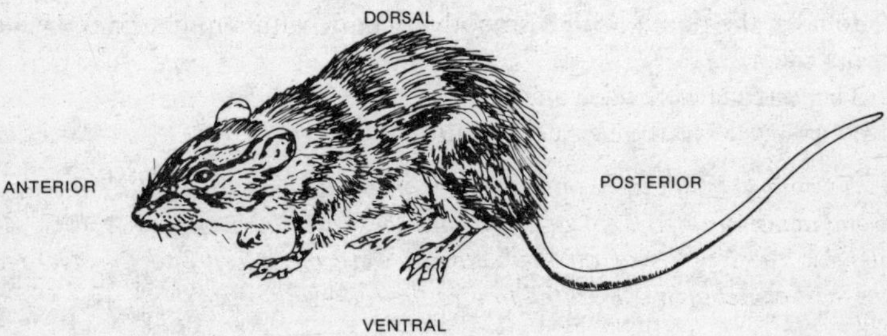

Fig. 2.3. *Orientation of the body*

When drawing a section, always give its orientation. A transverse section (T.S.) is really another name for a cross-section. In a structure which has a long and a short axis, a T.S. is a cut across the long axis. A longitudinal section (L.S.) is a cut along the long axis (see Fig. 2.2).

A vertical section (V.S.) is an upright section and is commonly used for describing sections through the kidney, heart and skin. It is particularly used for structures which do not have a noticeable long or short axis.

In animal drawings you should be aware which is the anterior (head) end, the posterior (tail) end, and the front (ventral) and back (dorsal) surface – see Fig. 2.3.

Graphs

You may have to construct a graph from data given to you, or you may be provided with a graph, and have to answer questions by interpreting or reading off the graph.

In drawing graphs, take note of the following points:

1. Plan how you are going to fit the graph on to the paper provided.
2. The horizontal or x-axis is used for the controlled variable. This means those units which are determined and controlled by the experimenter.
3. The vertical or y-axis is used for the uncontrolled variable. These are the results which the investigator finds out from the experiment.
4. Each axis must be divided into the correct number of units marked off at equal intervals.
5. Each axis must be fully and accurately labelled.
6. Plot each point with a circled dot or with a small cross.
7. Join up the points with a smooth curve or with straight lines drawn with a ruler.
8. The graph should have a title.

Let us try this out on an examination question.

Q. Twenty tadpoles were put into each of two aquaria on day 0. The water in one aquarium was kept at 28 °C and the water in the other at 7 °C. At intervals the lengths of all the tadpoles were measured and the average length of each batch calculated to give the following results:

	average length in mm	
day	tadpoles kept at 28 °C	tadpoles kept at 7 °C
0	*15*	*15*
1	*19*	*17*
6	*25*	*19*
9	*29*	*20*
13	*33*	*21*
16	*35*	*21*
20	*38*	*22*
23	*30*	*22*

(*a*) *Plot these results on the squared paper using one pair of axes.*

What we have here are two curves plotted on one set of axes.

Time (in days) is the controlled variable – it is determined by the experimenter. It goes on the horizontal axis. The average length of tadpoles (in mm) is the uncontrolled variable – it is what the experimenter is finding out. It goes on the vertical axis.

Now we can plot the graph as follows (for the time being ignore the arrowed broken lines):

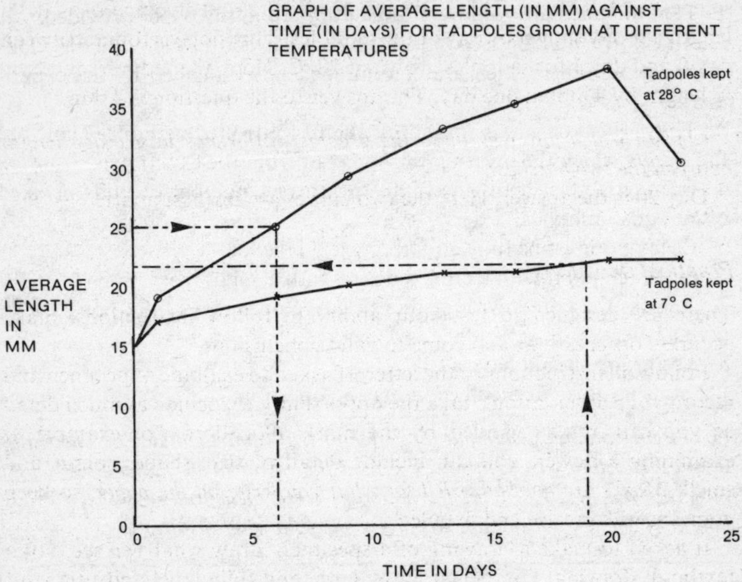

The question then goes on, with further sections asking us to interpret from the graph.

(*b*) *On day 19 what was the average length in mm of the tadpoles kept at 7°C?*

To do this, we find day 19 on the horizontal axis, draw a vertical line leading up to the appropriate curve, and then draw a horizontal line leading to the vertical axis. Here we read off the value, which is 21.7 mm.

We can easily do this the other way round, working from a value on the vertical axis to one on the horizontal axis. For example, if we are given the average length of tadpoles grown at a particular temperature, we can find out how long they have been in the aquarium. The tadpoles grown at 28 °C, and which have an average length of 25 mm, have been in the aquarium 6 days.

(*c*) *In any one day what was the greatest average growth in length shown by any of the tadpoles?*

In a graph, the slope of the curve indicates the rate of change. The steeper the slope, the faster the rate of change.

In our graph, we are looking for the part of the curve which has the steepest uphill slope. This will give us the greatest average growth in length per day shown by any of the tadpoles. This slope is found between day 0 and day 1 for tadpoles grown at 28 °C. Here, the tadpoles increase in length by 4 mm in one day. The answer to the question is 4 mm.

(*d*) *On which day was there the greatest difference in average length between the two batches?*

Day 20 is the answer. Here the two curves are farthest apart.

Practical questions

These are designed to test your ability to follow instructions, make accurate observations and come to valid conclusions.

Follow all instructions to the letter. If asked to examine a specimen and record your observations, take the opportunity to include as much detail as you can, but be guided by the mark allocation. For example, if examining a flower, you can include detail of size, shape, colour and smell. *Marks are only awarded for what you write on the paper*, so keep your answers precise and detailed.

If asked to make a drawing of a specimen, draw what *you* see, not a textbook drawing. Use pencil, draw large and fully label and title your drawing; include scale or magnification.

The following topics are popular in exams: flower structure, fruit and seed structure, vegetative organs (e.g. bulbs, corms and rhizomes, twigs), comparisons/drawings of bones or teeth, external features of small invertebrates, stages in life cycle of insects. Simple experiments may be set on topics such as osmosis, behaviour in small invertebrates, or using food tests.

3. *The Organization of Life*

To begin to appreciate what a living organism is, in a biological sense, you need to know what it is made of and what it is capable of doing. Despite the enormous range of life forms – from microscopic specks to oak trees and human beings – all living organisms are made up of similar substances and show certain characteristic features.

3.1. *The Chemistry of Life*

Organisms are made up of four common chemical elements – hydrogen (H), carbon (C), nitrogen (N) and oxygen (O). Seven other elements, sodium (Na), magnesium (Mg), phosphorus (P), sulphur (S), chlorine (Cl), potassium (K), and calcium (Ca), are present in moderate amounts. A number of elements, such as iron (Fe) and iodine (I), are also present in minute amounts.

These elements are arranged into inorganic and organic molecules. **Inorganic molecules** tend to be small and simple and consist of two or more elements chemically bonded together. Inorganic molecules are abundant in both living and non-living things. One of the commonest is water (H_2O).

Surprisingly enough, most organisms are made up of over two-thirds water. The chemistry of life is dominated by the chemistry of water.

1. Water is the **universal solvent**. It is the liquid in which substances in the body are dissolved, and all the body's chemical reactions take place in it.
2. Water itself takes part in certain chemical reactions, such as photosynthesis (page 48) and digestion (page 65).
3. It serves as a medium for the transport of materials (page 116).

Organic molecules are a characteristic feature of living, or once-living, organisms. These molecules are based on carbon and hydrogen to which oxygen and other elements are often chemically bonded. Such molecules tend to be complex and sometimes very large. In living organisms, four types are important: carbohydrates, proteins, fats and nucleic acids. These are dealt with in section 3.8.

3.2. *The Characteristics of Life*

All living organisms demonstrate *all* seven of the following characteristics at some time in their life.

1. **Movement**: all living things move either a part or the whole of their bodies (Chapter 10).
2. **Feeding**: living things feed. Food is the material from which organisms obtain (i) the energy, and (ii), the raw material, to construct, repair and maintain the body. The study of food and feeding processes is termed **nutrition** (Chapter 5).
3. **Respiration** is a chemical reaction, or rather a series of chemical reactions, by which organisms release energy from organic foods (Chapter 6).
4. **Excretion** is the removal of waste substances produced by chemical reactions in the body (Chapter 8).
5. **Growth**: living organisms grow, i.e. increase in size. The term growth is usually taken to include increase in complexity as well as size (Chapter 12).
6. **Reproduction** is the production of new individuals (offspring) derived from existing individuals (parents) (Chapter 11).
7. **Sensitivity**: living organisms are sensitive. They have the ability to respond to a stimulus (Chapter 9).

A convenient way to remember these seven characteristics is to use a mnemonic such as:

Many	**M**ovement
Foolish	**F**eeding (nutrition)
Rabbits	**R**espiration
Eat	**E**xcretion
Green	**G**rowth
Rhubarb	**R**eproduction
Shoots	**S**ensitivity

These features are not necessarily very easy to demonstrate. In humans, for example, respiration is accompanied by obvious breathing movements (see page 104). Plants, however, do not breathe and complex apparatus is required to demonstrate respiration (see page 94).

Non-living objects may show certain characteristics of living organisms, but not all of them. A motor car takes in oxygen and gives out carbon dioxide (respires), it consumes fuel (feeds), it releases exhaust fumes (excretes), and it moves. Arguably, it is also sensitive (in a

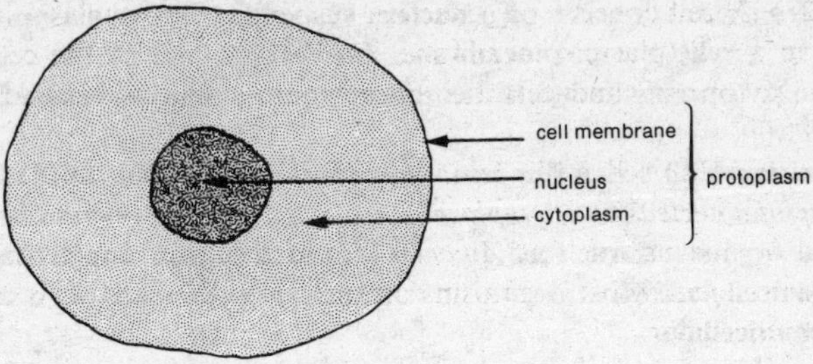

Fig. 3.1. *The chief characteristics of cells*

biological sense). If you switch on the ignition the car will (if you are lucky) respond by the engine starting. What a car does not do, however, is grow or reproduce. It does not fulfil all the seven characteristics of life.

3.3. *Metabolism*

Living things are complex arrangements of substances not normally found in non-living situations. To make complex substances from simpler ones requires energy. Organisms thus need a continual supply of energy to make the organic substances from which they are constructed. The energy is used in forming chemical bonds within these molecules.

Metabolism is a word used to describe all the chemical reactions which occur in an organism and are necessary for life. There are two types of metabolism: **catabolism** and **anabolism**.

Catabolism is the process by which complex substances are broken down into simpler ones, resulting in the release of energy. Respiration (page 91) is an important example of catabolism.

Anabolism is the process by which simple substances are built up into complex compounds. This type of reaction requires a supply of energy. Examples of anabolism are photosynthesis (page 48) and all cases of growth and repair in the bodies of organisms.

Both catabolic (breaking down) and anabolic (building up) processes occur in living things. Energy released from catabolic reactions is used to drive anabolic reactions.

3.4. *Cells*

Organisms are made up of cells, though there are some exceptions to this (page 27). A cell consists of a **nucleus** suspended in **cytoplasm** and enclosed in a **cell** (plasma) **membrane**. The living parts of the cell – the nucleus, cytoplasm and cell membrane – are sometimes called **protoplasm**.

Definition: The cell *is the basic unit of life. It is a unit of protoplasm made up of a nucleus and cytoplasm surrounded by a cell membrane.*

Some organisms, such as *Amoeba*, are made up of one cell and are called **unicellular**. Most organisms are made up of many cells and are called **multicellular**.

Why cells are small

Movement of substances in and out of cells is mainly by diffusion (see Chapter 4), which occurs most efficiently over short distances. The size of cells means that every part of the cell is close to its surroundings. Larger cells, such as birds' eggs and some plant cells, contain large amounts of non-living matter which do not need to obtain substances by diffusion.

The structure of a cell is closely related to its function. A nerve cell (**neurone**) in an animal has to transmit a nerve impulse from one place to another, often over a long distance. These neurones, therefore, are long and thin to perform their function.

3.5. *Plants and Animals*

Living organisms are usually separated into two major groups (kingdoms), plants and animals. There are a number of basic differences between plants and animals with which you should be familiar. These are summarized in Table 3.1.

3.6. *Cell Structure*

When viewed through a light microscope (which magnifies up to 1,500 times), only a few distinct structures are visible within the cell. Staining helps to reveal the nucleus, grainy cytoplasm and cell membrane. When examined with an electron microscope (which magnifies up to 500,000 times) many more structures become visible. The cytoplasm ceases to be a grainy mass, but is seen to be made up of compartments. These compartments, together with the nucleus, are called **organelles**. They have different shapes and sizes and perform different functions.

Definition: an organelle *is a structure within the cell which has a special function or functions, e.g. the nucleus.*

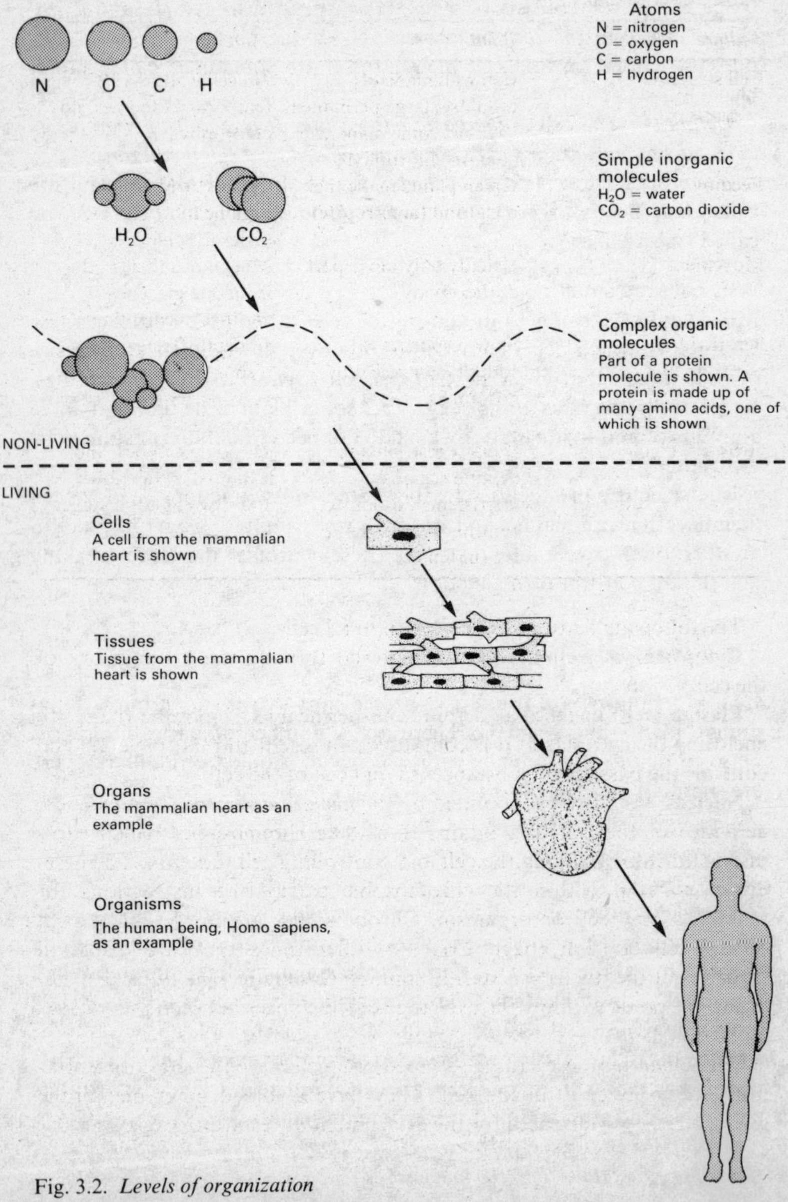

Fig. 3.2. *Levels of organization*

Table 3.1. *The main differences between plants and animals*

Feature	*Plant*	*Animal*
Cell structure	Cell wall made of cellulose, large permanent vacuole, some plant cells contain chlorophyll	No cell wall small, temporary vacuoles, no chlorophyll
Feeding	Green plants make their own food (**autotrophic**)	Obtain 'ready-made' organic food (**heterotrophic**)
Movement	Usually only move part of their body	Most move their body from one place to another (**locomotion**)
Sensitivity	Slow responses to stimuli, controlled by hormones	Generally faster responses, controlled by nervous system as well as hormones
Growth	Usually restricted to definite regions (meristems), usually continues throughout life (unlimited).	Throughout body and is limited, slows down and stops at adult stage

The following features are common to all cells:

Cytoplasm: a grainy, jelly-like material forming the groundwork of the cell.

Plasma (cell) **membrane**: a thin semi-permeable membrane (page 40) enclosing the cytoplasm. It is composed of protein and fat (page 29) and controls the passage of substances in and out of the cell.

Nucleus: the nucleus, bounded by the **nuclear membrane**, controls the activities of the cell. It contains threadlike **chromosomes** which carry instructions for building the cell and controlling cell function. The **hereditary material** within the chromosome carries the instructions for making a new cell or organism. Chromosomes are not visible except during cell division (page 251). At other times they are dispersed throughout the nucleus as darkly staining **chromatin**. The nuclear membrane has pores to allow the exchange of substances between the nucleus and cytoplasm.

Mitochondria: these sausage-shaped organelles are the sites of respiration (energy release) in the cell. They have a folded inner membrane giving a large surface area for the attachment of respiratory enzymes.

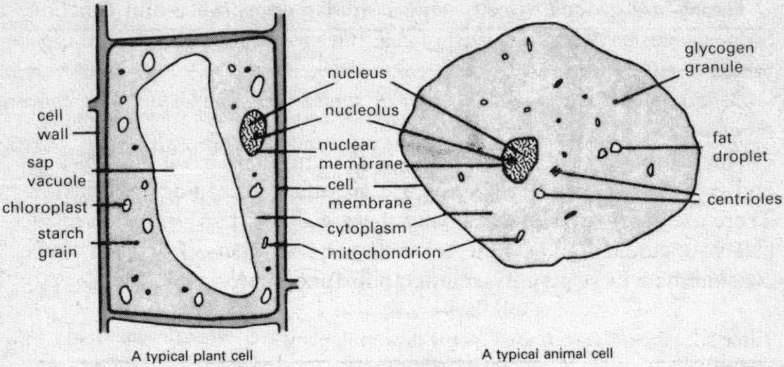

Fig. 3.3. *Comparison between a plant and an animal cell*

The following features are present only in plant cells (see Fig. 3.3.);

Cell wall: a rigid wall made of cellulose and surrounding the cell membrane. It gives shape and some support to the cell.

Vacuole: a large, non-living, fluid-filled space in the cytoplasm. It pushes the cytoplasm out against the cell wall and makes the cell turgid (page 43).

Chloroplasts: similar in size to mitochondria, these are the sites of photosynthesis in plant cells. They contain the green pigment **chlorophyll** together with the enzymes needed for the reactions of photosynthesis. (Note: not all plant cells contain chloroplasts.)

Starch: this is the storage carbohydrate in plant cells. It is found as grains in the cytoplasm.

The following features are found only in animal cells (see Fig. 3.3); those marked with an asterisk (*) are visible only with an electron microscope:

Centrioles*: during cell division, these cylindrical structures give rise to spindle fibres.

Glycogen: in animal cells glycogen is the storage carbohydrate. It is found as granules in the cytoplasm.

3.7. *Cells, Tissues and Organs*

All larger animals and plants are multicellular (many-celled). In these organisms the cells are grouped together to form tissues and organs.

Tissues are groups of cells with a similar appearance and function. Some tissues have just one type of cell, but most have two or three types mixed together.

Definition: a tissue *is a collection of similar cells performing the same function, e.g. muscle.*

The main types of tissue in plants and animals are listed in Tables 3.2 and 3.3. Notice that the tissues are classified according to function. There are other ways of classifying them. Rather than simply learn the list, you should look up the relevant page numbers. For each tissue consider how its structure is related to its function.

Table 3.2. *Plant tissues* (based on the flowering plant)

Name of Tissue	*Functions*	*Examples*
Protective tissue	Protection (also against desiccation)	Epidermis (page 309) Cork – dead cells (page 309)
Packing (supporting) tissue	Fills in space between other tissues and cell turgidity provides support	Parenchyma (page 203)
Photosynthetic tissue	Photosynthesis	Mesophyll (page 50)
Vascular tissue	Transporting water and food substances	Xylem – dead cells (page 136); phloem (page 145)
Strengthening tissue	Supporting the plant	Collenchyma; sclerenchyma – dead cells (page 203)
Meristematic (dividing) tissue	Cell division. Gives rise to other tissue types and enables growth	Shoot and root tips; cambium (page 309)

Organs are groups of tissues which combine together to form a structure which has a particular function or functions. An example is the mammalian heart (page 125). It contains epithelial tissue, cardiac muscle, nervous tissue and blood, all organized into a complex structure. One of the heart's functions is to pump blood around the body.

Definition: an organ *is a structure formed by two or more tissues working together to perform particular functions, e.g. the heart.*

Table 3.3. *Animal tissues* (based on the human)

Name of Tissue	*Functions*	*Examples*
Epithelial tissue	Lines and protects tubes and cavities within the body. Also protects body surface	Epidermis of skin (page 157); lining layer of the gut (page 76)
Connective tissue: has a matrix (non-living part)	Connects, binds, supports and protects (blood and lymph are exceptions)	Fibrous connective tissues such as tendons and ligaments (page 191); skeletal tissues – bone and cartilage (page 191); blood – a fluid tissue (page 117); lymph – a fluid tissue (page 129)
Muscular tissue	Brings about movement	Smooth (visceral) muscle; striped (skeletal) muscle; cardiac (heart) muscle (page 197)
Nervous tissue	Communication – transmits nerve impulses	Conducting tissue: motor, sensory and relay (page 164); non-conducting tissue: myelin sheath

In flowering plants (page 308) there are four major organs – the stem, root, leaves and, at certain times, the flower. The leaf, for example, is made up of epidermal, photosynthetic, packing, vascular and strengthening tissues (see pages 25 and 50). The leaf's main function is photosynthesis (page 59).

In the more complex animals such as insects (page 320) and mammals (page 329) the organs do not operate independently but work together as **organ systems** to take on the responsibility for particular functions. For example, in the human, the gut, liver and pancreas work together to digest and absorb food. They form the digestive system (page 73).

Figure 3.2 (page 22) attempts to show the relationship between chemical substances, cells, tissues and organs, and the organism as a whole. Clearly, not all organisms show all these levels of organization. Unicellular organisms, such as *Amoeba* (page 315) and *Chlamydomonas* (page 297), are not made up of tissues or organs; all life processes are carried out within a single cell. Some simpler multicellular organisms such as the coelenterate, *Hydra*, function at the tissue level of organization; life processes are carried out mainly by tissues and isolated cells.

A number of organisms are not made up of cells as we defined them earlier:

1. The non-green plants, **fungi** (page 301), have tubular structures called hyphae which contain cytoplasm. Within these are found many nuclei. No membrane or cell wall separates off one nucleus and region of cytoplasm from another to form recognizable cells.
2. **Bacteria** (page 294) are unicellular but do not have a proper nucleus. The genetic material is free in the cytoplasm and is not enclosed in a nuclear membrane.
3. **Viruses** (page 292) are on the borderline between living and non-living. They are extremely small (visible only with an electron microscope) and have a structure unlike any other kind of organism. They have no plasma membrane, nucleus or 'cytoplasm'.

3.8. *Biological Molecules* *

You need to know the structure and function of the major types of biological molecule. Without this, you will not properly appreciate biological processes such as nutrition, respiration or excretion.

Carbohydrates

These contain the elements carbon, hydrogen and oxygen.

Carbohydrates are made up of units called six-carbon sugars. These units can be joined together or broken apart, rather like beads on a string. **Monosaccharides** are made up of one unit, **disaccharides** are made of two, and **polysaccharides** are made up of many.

Soluble carbohydrates are called sugars. In both animals and plants, simple sugars can be built up into complex carbohydrates and then broken down again as required. Carbohydrates are transported round the body in solution as sugars and are stored within cells as complex insoluble molecules. Being insoluble, these large storage molecules do not produce an osmotic potential (see page 42) and so do not create problems of water movement. The joining together of sugar units is a **condensation** reaction. A molecule of water is released when each chemical

* Check with your syllabus how much detail is required.

	Chemical formula	*Solubility*	*Examples*
Simple sugars (monosaccharides): one sugar unit o	$C_6H_{12}O_6$	Soluble	Glucose Fructose Galactose
More complex sugars (disaccharides): two sugar units o—o	$C_{12}H_{22}O_{11}$	Soluble	Maltose (milk sugar) Sucrose (table sugar)
Large molecule carbohydrates (polysaccharides): many sugar units o—o—o—o—o—o—o	$(C_{12}H_{22}O_{11})_n$	Insoluble	Starch (in plants) Cellulose (in plants) Glycogen (in animals)

bond between two sugar units is formed. The breaking apart of sugar units is a **hydrolysis** reaction. A molecule of water is used in breaking each bond between units. Digestion (page 65) is a hydrolysis reaction.

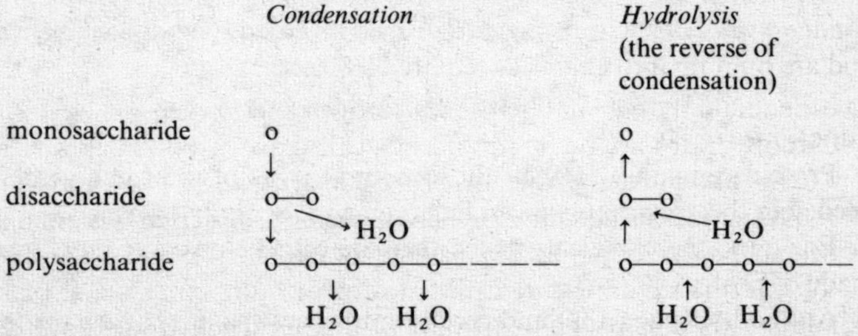

The building up and breaking down of carbohydrates

Functions

1. Monosaccharides are the main substances broken down in respiration (page 92) to release energy.
2. Polysaccharides are used as energy stores – in plants, **starch**; in animals, **glycogen**.
3. The polysaccharide **cellulose** is used by plants as structural material in the cell wall.

Proteins

Proteins contain the elements carbon, hydrogen, oxygen, nitrogen and often sulphur.

In the same way that polysaccharides are made up of chains of sugar units, so proteins are made up of chains of amino acids. However, whereas there is a choice of only one or a few different types of sugar unit in a carbohydrate, in a protein there are up to 20 different types of amino acid. These can be joined together in any order to form chains hundreds of units long, called **polypeptides**. The arrangement, or sequence, of amino acids in the chain varies from one protein to another. For example, in one protein

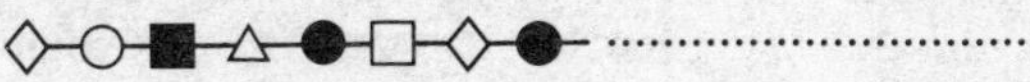

and in another

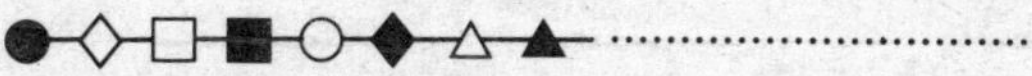

where each symbol represents a different amino acid.

Just as with polysaccharides, proteins can be broken down by hydrolysis into their constituent units (amino acids) and then built up again by condensation. They are transported round the body as the single units and are built up into chains as required by cells.

Functions

1. Proteins make up most of the structural parts of the cell and so are needed in the growth, repair and maintenance of all living cells.
2. Enzymes, the chemicals which regulate cell chemistry (page 33), are made of protein.
3. Antibodies (page 132) and certain hormones (page 176) are made of protein.

The immense variety of life is largely due to the enormous number of different proteins that exist.

Fats

Fats, like carbohydrates, contain only carbon, hydrogen and oxygen. However, in fats, the proportion of oxygen is lower. A typical fat is tristearin, $C_{57}H_{110}O_6$.

The most common type of fat has four parts, one part derived from glycerol and three parts from fatty acids.

A fat molecule

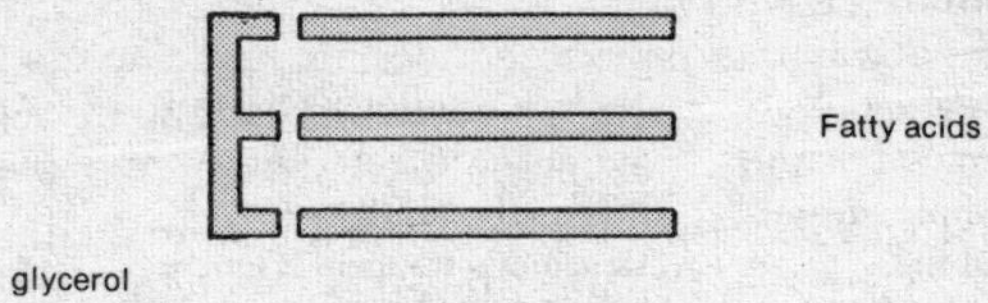

When a fat molecule is broken down by hydrolysis, the glycerol part is separated from the three fatty acids. The original fat, or a different one, can be built up by condensation.

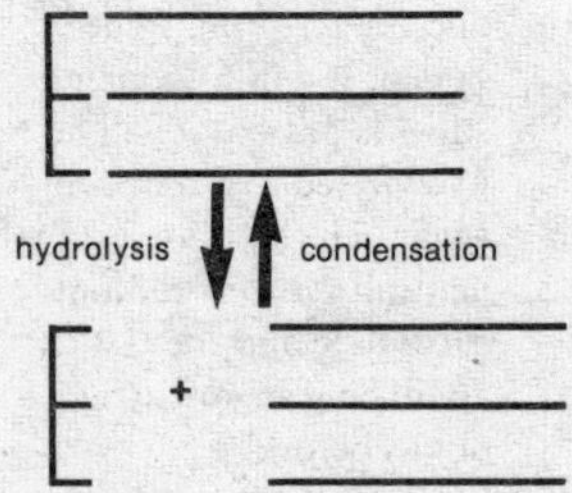

Functions

1. They form part of the membranes of cells.
2. They are an important energy store and in respiration (page 92) release twice as much energy per gram as carbohydrates or proteins.
3. In mammals, fat stores protect internal organs from damage and fat under the skin acts as heat insulation.

Carbohydrates, **proteins** and **fats** are classed as foods (see page 67). Their presence in organic material can be determined using food tests (see Table 3.4). Questions on these tests are extremely popular in exams.

Testing for vitamin C

This test depends on the ability of vitamin C (ascorbic acid) to decolourize a blue dye called DCPIP. The method described below explains how to find the vitamin C content of a test sample.

1. Put 1 cm^3 of DCPIP (2,6-dichlorophenol indophenol) solution into each of two test tubes.
2. Draw up 2 cm^3 of 0.1 per cent vitamin C solution into a syringe.

Table 3.4. *Food tests*

Food	*Test*	*Positive result*
1. CARBOHYDRATES		
(a) *Reducing sugars*: monosaccharides (simple sugars) and the disaccharide maltose	Make a solution of the test substance in water. Add an equal quantity of **Benedict's Reagent**. Gently bring to the boil	An orange or green precipitate
(b) *Non-reducing sugars*: disaccharides other than maltose (NOTE: this test should only be used after test (a) has given a negative result)	Dilute hydrochloric acid is added to the test sample and the resulting solution boiled. If a non-reducing sugar is present it will be hydrolysed to a reducing sugar. The solution is then neutralized with sodium carbonate and the test for a reducing sugar carried out as before	An orange or green precipitate
(c) *Polysaccharides*	Add iodine solution to the test substance	For starch a blue-black colour forms; for cellulose: blue; for glycogen: reddish-brown
2. PROTEINS	Use either of these tests:	
	1. Place test substance in a test tube. Add twice the volume of **Millon's Reagent** and heat gently	The test substance turns red on warming
	2. **Biuret test** for dissolved proteins. To the test solution add an equal volume of sodium hydroxide and one or two drops of copper sulphate solution	On shaking, a mauve colour appears

Food	*Test*	*Positive result*
3. FATS AND OILS	Use any of the following tests:	
	1. Rub test solid on to filter paper	Grease spot forms on paper
	2. Shake test substance with ethanol. Pour ethanol into water	Milky suspension
	3. Add few drops of red dye Sudan III to the test substance in water. Shake. Allow to stand.	Red oil layer floats on surface

3. Add this vitamin C, drop by drop, to the first test tube.
4. Note the volume required to turn the DCPIP colourless.
5. Repeat the procedure using your test sample.
6. Note the volume of the test sample that was needed to decolourize the DCPIP.

A comparison of these two volumes can now be used to calculate the amount of vitamin C in your sample.

For example, let us assume that it takes 0.6 cm^3 of 0.1 per cent vitamin C and 1.2 cm^3 of the test material to decolourize equal volumes of DCPIP.

As it takes twice as much of the sample to make the DCPIP colourless your sample contains half the amount of vitamin C as was in the pure solution. Therefore your sample contains 0.05 per cent vitamin C.

Q. For each of the following food materials describe one test you would use to show its presence:

(*i*) *a reducing sugar;*
(*ii*) *starch;*
(*iii*) *a protein;*
(*iv*) *a fat.*

3.9. *Enzymes*

Enzymes are 'biological catalysts'. Catalysts are chemicals which speed up chemical reactions. The catalyst itself remains unchanged by the reaction.

Unlike the inorganic catalysts used in the chemical industry, enzymes are made of protein. Without enzymes, chemical reactions in the body would be too slow to support life.

Enzymes, just like other proteins, are manufactured within cells. The properties of enzymes can be summarized as follows:

1. They are proteins made by living cells (they sometimes incorporate vitamins).
2. They are catalysts.
3. They are specific, i.e. one type of enzyme catalyses one type of reaction.
4. An enzyme works best at a particular temperature (the optimum temperature). As with other proteins, they are destroyed (denatured) by heating.
5. Each enzyme works best at a certain degree of acidity or alkalinity (the optimum pH).

Outside its optimum temperature and pH an enzyme may cease to work or may even be destroyed (we normally say '**denatured**' to indicate its chemical nature has been lost). The optimum temperature for most enzymes is the human body temperature, 37 °C. The optimum pH for enzymes varies. For pepsin (page 75) it is pH 2 (very acid), and for salivary amylase (page 74) pH 7.5 (nearly neutral).

Often the name of an enzyme ends in **-ase**. The first part of the name may indicate the substrate (reactant) whose reaction the enzyme catalyses. For example, **proteases** catalyse the breakdown of proteins; **carbohydrases**, the breakdown of carbohydrates; and **lipase**, the breakdown of lipids. Individual enzymes, however, often do not follow this rule. **Pepsin** (page 75) is an example of a protease. **Salivary amylase** (page 74) is a carbohydrase which hydrolyses (digests) starch.

Enzymes catalyse *all* chemical reactions in the body. However, the enzymes you are probably most familiar with are digestive enzymes. These are secreted into the gut and are, in fact, unusual enzymes because they are **extracellular** (work outside cells). Most enzymes are not secreted but are **intracellular** (work inside cells). One such enzyme in plants is **starch phosphorylase** found in the cotyledons of bean seeds (page 221). It catalyses the first reaction in the conversion of starch to sugar and is therefore important in germination, when stored carbohydrates are being utilized. An example of an intracellular enzyme in animals is **carbonic anhydrase** (page 117) found in the red blood cells of mammals.

Questions on the effect of temperature and pH on enzyme activity, together with experiments, are very popular with examiners. For example:

Q. (*a*) *What are the effects of a change in*
(*i*) *temperature*
(*ii*) *pH upon the rate of action of any* named *enzyme*
(*b*) *Describe with full experimental details how you would investigate either of these effects.*

In such questions it is usual to use a digestive enzyme as the named example, in the experiment, and to use a food test (page 31) to test for the presence of substrate or product.

Investigating the effect of temperature on the digestion of starch by salivary amylase (ptyalin)

This experiment investigates how rapidly starch is digested at different temperatures under the reaction of salivary amylase. When a starch solution no longer gives a positive test with iodine (see page 31) it can be assumed that the starch has been digested.

Method

1. A 1 per cent starch solution is made up and 5 cm^3 added to each of five test tubes.
2. 1 cm^3 of freshly collected saliva is added to each of four other test tubes.
3. The test tubes are placed in water baths at various temperatures as shown.

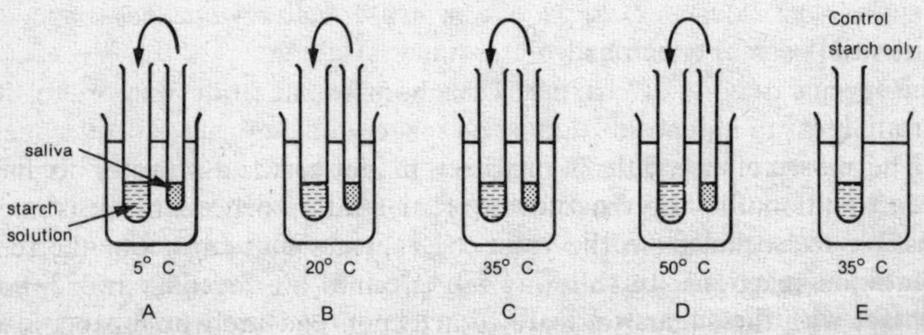

Fig. 3.4. *Apparatus for investigating the effect of temperature on starch digestion*

4. The test tubes are left in their respective water baths to reach the required temperature, and then the solutions in the two tubes are mixed.
5. At timed one-minute intervals a drop of each mixture is removed by dropper pipette and tested for starch, using iodine. NOTE: iodine is *not* added direct to the mixture – this would affect the reaction.
6. For each mixture, the time is recorded when a drop of the solution no longer turns the iodine blue-black, i.e. starch has been digested. The experiment is stopped after 20 minutes.

Results

Typical results are as follows:

A. Starch test still positive after 20 minutes.
B. Starch test negative after 8 minutes.
C. Starch test negative after 2 minutes.
D. Starch test still positive after 20 minutes.
E. Starch test still positive after 20 minutes.

If mixtures A and D are then transferred to 35 °C for 20 minutes, A will give a negative test for starch and D a positive test.

Conclusions

The results taken as a whole indicate that the enzyme works best (digests starch fastest) at 35 °C, and at a lower temperature the reaction is slower. At 50 °C, the enzyme is denatured. This is indicated by the enzyme not working when kept at 35 °C. The control shows that starch does not break down of its own accord.

Here are two final points regarding enzymes:
1. Students sometimes write of enzymes being 'killed'. Enzymes cannot be killed, they are non-living. They can, however, be denatured.
2. Enzymes catalyse *all* the chemical reactions of metabolism, not just those concerned with digestion.

Industrial uses of enzymes

The specific actions of enzymes have been found to be very useful in certain areas of industry.

The most common use is probably in the brewing industry. It has been traditional to use yeast for the fermentation process but it is now possible to extract from the yeast the enzymes that carry out the fermentation reactions. In this way the enzymes are brought into direct contact with the sugars they are to act upon and the whole process is speeded up.

Another use of enzymes is in genetic engineering, where they are used to separate sections of the genetic material in cells. These genes can then be transferred to other cells. A similar process is used in the manufacture of certain hormones required for medical treatment. Enzymes are used to modify the structure of plant hormones to make them similar to animal hormones.

There are many other ways in which enzymes are useful in industry and they are likely to increase in importance in the future.

Definitions

Cell
Organelle
Tissue
Organ

Key Words

Inorganic molecules	Anabolism	Cell wall
Organic molecules	Nucleus	Vacuole
Universal solvent	Cytoplasm	Chloroplasts
Carbohydrates	Cell(plasma) membrane	Starch
Proteins	Unicellular	Centrioles
Fats	Multicellular	Glycogen
Nucleic acids	Phagocytic	Monosaccharides
Movement	Autotrophic	Disaccharides
Feeding	Photosynthesis	Polysaccharides
Nutrition	Heterotrophic	Sugars
Respiration	Locomotion	Hydrolysis
Excretion	Unlimited growth	Condensation
Growth	Nuclear membrane	Amino acids
Reproduction	Chromosome	Fatty acids
Sensitivity	Hereditary material	Glycerol
Metabolism	Chromatin	
Catabolism	Mitochondria	

Questions

1. (a) Make a labelled diagram to show the structure of a named unicellular animal.

(b) How does the structure of this animal differ from:

(i) a simple animal cell such as from the lining of your cheek;

(ii) a named unicellular plant?

(c) For each of the following organelles: mitochondrion, chloroplast, nucleus, plasma membrane, vacuole, give one function.

2. What is meant by the terms

(i) carbohydrate

(ii) fat

(iii) protein?

3. (a) What are enzymes?

(b) What are the main properties of enzymes?

(c) Describe, with full practical details, an experiment you would perform to find out how the rate of an enzyme-controlled reaction is related to temperature.

(d) What results would you expect to obtain from your experiment and how would you express them?

4. $C_{12}H_{22}O_{11}$ *is the chemical formula for*

A. an amino acid

B. a protein

C. a fat

D. a carbohydrate

E. a nucleic acid.

4. *Movement of Substances Into and Out of Cells*

Substances generally move into and out of cells by one of three processes: diffusion, osmosis or active transport. These processes are a popular topic with examiners and a clear understanding of them is essential.

As we have seen in Chapter 3, living organisms are largely made up of water. All living cells are surrounded by water and any substance that passes into or out of a cell must be in solution.

4.1. *Diffusion*

The molecules in a liquid or gas are constantly moving. Unless prevented from doing so by a barrier, they tend to move from a region where they are at a high concentration to a region where they are at a lower concentration. This movement continues until the molecules are evenly distributed. For example, if you drop a small amount of liquid dye into a container of water, the dye molecules will eventually spread throughout the container colouring the whole water. The dye molecules have spread from a region of high concentration to a region of low (initially zero) concentration. This movement of molecules is a physical process termed diffusion.

Definition: Diffusion *is the free movement of molecules of a substance from regions of high concentration to regions of lower concentration. The process continues until the molecules are evenly distributed.*

In an organism, many substances are in solution and diffusion is the usual way they move into, out of or between cells.

Examples of diffusion in living cells are:

1. Diffusion of oxygen and carbon dioxide (gas exchange) across the lining of alveoli in the lungs of mammals (see page 109).
2. Diffusion of oxygen and carbon dioxide (gas exchange) through the stomata of leaves in higher plants (see page 113).

In fact, wherever gas exchange occurs, diffusion of oxygen and carbon dioxide is occurring (for other examples see page 99–104).

Any substance that can pass through a cell membrane will diffuse into or out of a cell if its concentration inside the cell is different to that outside.

As an illustration of this, let us look at a smooth muscle cell in a mammal. As the muscle cell respires it uses up oxygen and the oxygen concentration in the cell decreases. The concentration of oxygen outside the cell is higher so that oxygen will diffuse into the cell.

As the cell respires, it gives off carbon dioxide. The concentration of carbon dioxide in the cell will rise above that of its surroundings and carbon dioxide will diffuse out of the cell.

Glucose, another substance used up in cell respiration, will diffuse into the cell to replace glucose that has been broken down.

Think of any cell you like and try and work out what chemical processes are going on inside it. For example, in a mesophyll cell in the leaf of a green plant exposed to sunlight, photosynthesis as well as respiration is occurring (page 62). What substances are being used up by the cell and what substances are given off? By asking these questions you should be able to work out what substances are diffusing into or out of the cell and in what direction.

To return to our muscle cell, we can summarize the diffusion of substances into and out of the cell in a diagram:

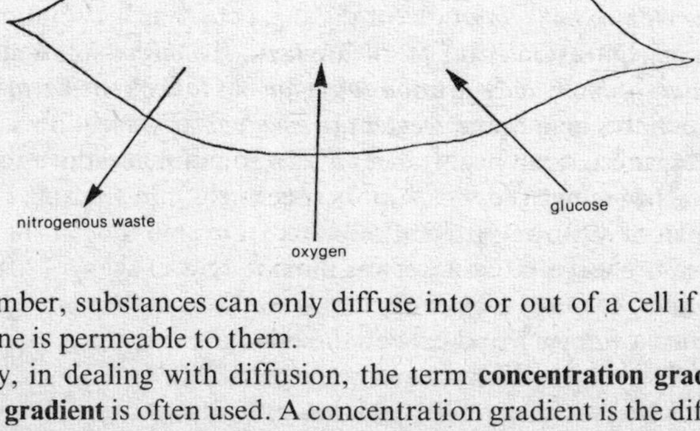

Remember, substances can only diffuse into or out of a cell if the cell membrane is permeable to them.

Finally, in dealing with diffusion, the term **concentration gradient** or **diffusion gradient** is often used. A concentration gradient is the difference in concentration between one region and another. Substances diffuse along a concentration gradient from a high to a low concentration. The greater (steeper) the concentration gradient the faster the rate of diffusion. Surfaces where gas exchange occurs are often designed to maintain a steep diffusion gradient so that diffusion occurs rapidly (see pages 99–111).

4.2. *Osmosis*

Osmosis is a special type of diffusion. It is the diffusion of water across a semi-permeable (sometimes known as a selectively permeable) membrane. Cell membranes are semi-permeable and it is by osmosis that water moves into, out of, and between cells. A semi- (selectively) permeable membrane is one that allows small molecules, e.g. gases and water, to pass across it, but not larger molecules, e.g. sugars and proteins.

Let us see how osmosis works. Look at Fig. 4.1 which shows a simple apparatus used to demonstrate osmosis. Inside the thistle funnel is a strong solution of sugar while outside in the beaker is pure water. Stretched across the bottom of the funnel separating the sugar solution from the pure water is a semi-permeable membrane. This can be a natural membrane, e.g. a pig's bladder, or an artificial one, e.g. cellophane. Let us see what happens.

After a time the level of liquid in the thistle funnel will rise. Water has flowed by osmosis from the pure water, across the semi-permeable membrane and into the sugar solution. Why is this so?

What the water has done is simply diffuse along a concentration gradient from a high concentration of water (the pure water) to a low concentration of water (the sugar solution). Put another way, water is diffusing from a weak solution (of the sugar) to a strong solution.

The sugar molecules cannot diffuse from the sugar solution into the pure water because they are too large to pass through the membrane. Remember the membrane is semi-permeable, it will allow the water molecules through but not the sugar. Water molecules move through in both directions, but the water molecules move into the sugar solution faster than they move out. Thus there is an overall movement of water from the pure water into the sugar solution.

Definition: Osmosis *is the diffusion of water molecules through a semi-permeable membrane from a weak solution to a strong solution.*

You must be absolutely clear about this definition. Osmosis involves **water** diffusing through a **semi-permeable** membrane.

Let us return to the apparatus in Fig. 4.1. We could apply an external pressure to the sugar solution in the thistle funnel; a pressure that would

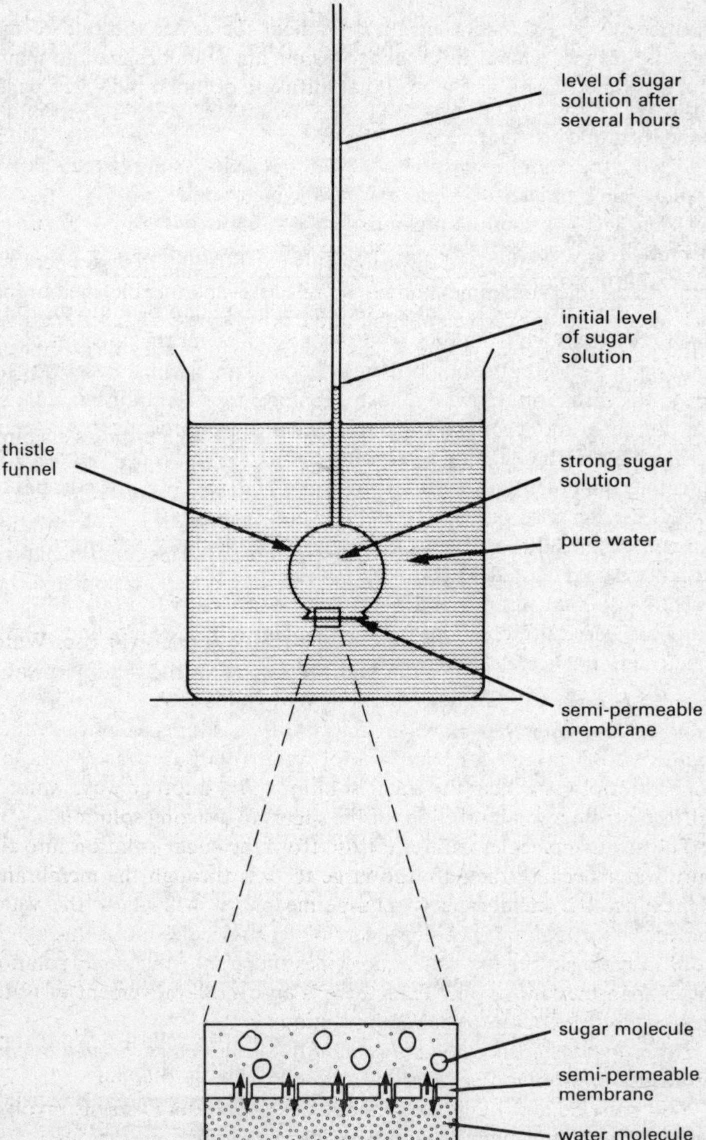

Fig. 4.1. *Demonstration of osmosis*

be just sufficient to stop the flow of water into the sugar solution. This pressure is called the osmotic pressure of the solution. The stronger (more concentrated) the solution, the higher its osmotic pressure. Thus, water will diffuse from a solution of low osmotic pressure to one of higher osmotic pressure.

Strictly, the osmotic pressure of a solution is always measured relative to pure water and across a semi-permeable membrane.

Definition: The osmotic pressure *of a solution is the pressure required to prevent the movement of pure water into the solution across a semi-permeable membrane.*

Really, we can only talk of a solution as having an osmotic pressure when it is separated from pure water by a semi-permeable membrane. A solution in a glass bottle or a cell in the body of an organism is obviously not in this situation. However, such solutions are potentially capable of exerting an osmotic pressure if separated from pure water by a semi-permeable membrane. For such solutions we say they have an **osmotic potential**, the potential to exert an osmotic pressure. Strictly, we should always use the term osmotic potential when dealing with body fluids or cell contents inside an organism. Throughout this book we will use the term osmotic potential (O.P.).

The ways that animal and plant cells respond to osmotic flows of water are very different. This is because plant cells are surrounded by a rigid cell wall whereas animal cells are not.

Osmosis and animal cells

The human red blood cell is a convenient animal cell in which to investigate osmosis. The osmotic potential of a red blood cell, as with other cells, is determined by the concentration of dissolved salts and dissolved organic substances it contains.

If a red blood cell is put in water, water will enter the cell by osmosis and the cell will swell up and may eventually burst.

If a red blood cell is put in a solution of greater concentration than itself, e.g. a strong salt solution, then the cell will lose water by osmosis and will shrink and wrinkle (Fig. 4.2).

Swelling up or shrinking can be damaging to a cell. To avoid this, animal cells are either surrounded by a liquid of the same osmotic potential as their own cell contents, or the cells have special mechanisms to regulate water content.

Multicellular animals maintain their body fluids at a relatively constant O.P. using a variety of methods: vertebrates do so using kidneys

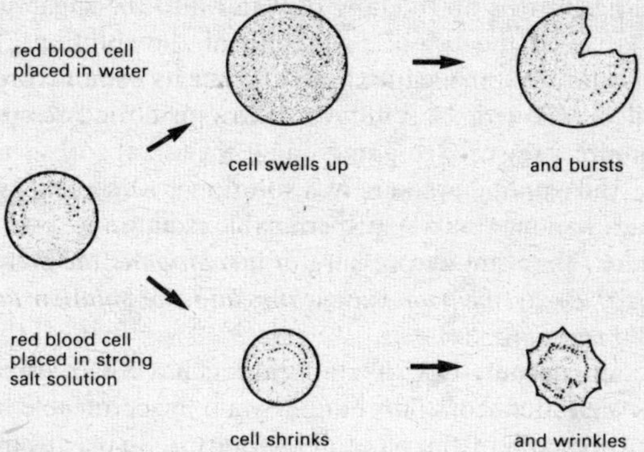

Fig. 4.2. *The effect on a red blood cell of changing the composition of the surrounding liquid*

(pages 151–56). Single-celled animals, on the other hand, are able to actively pump water out using contractile vacuoles (page 150).

All these methods of regulating osmotic potential come under the term **osmoregulation** (page 149).

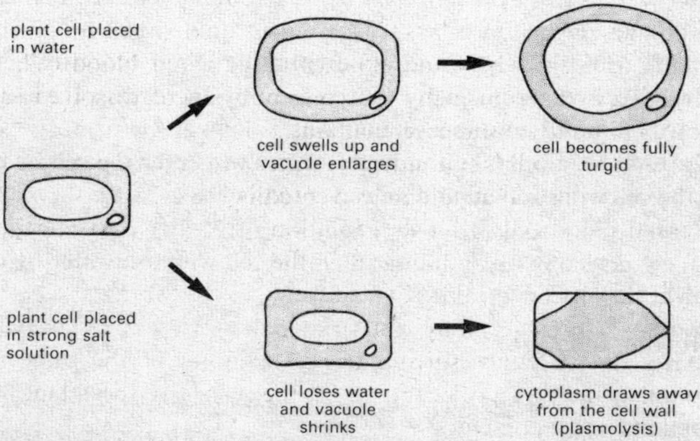

Fig. 4.3. *The effect on a plant cell of changing the composition of the surrounding liquid*

Osmosis and plant cells

A plant cell has a cell wall as well as a cell membrane. The cell wall is fully permeable (allows dissolved solids through as well as water) and it is rigid. It prevents the cell from bursting if the cell absorbs water.

If a plant cell is put in water, water enters by osmosis and the cell swells up but does not burst. The cell wall stretches slightly but does not break. At this point the cell is turgid. In general, plant cells are slightly turgid. Turgidity, or turgor, plays an important role in support in herbaceous (non-woody) plants (see page 203). When all the cells in a particular plant tissue absorb water, each cell swells and presses out against its neighbours. The result is a rigid mass of cells which can support the soft parts of the plant. The packing tissue, parenchyma, in particular plays this role.

If a plant cell is put in a strong sugar solution water passes out of the cell by osmosis and the cell loses its turgor and becomes flaccid. Under experimental conditions, if the external solution is strong enough, the cytoplasm shrinks and eventually pulls away from the cell wall, a process called plasmolysis. Plasmolysis does not normally occur in a living plant. A loss of turgidity, however, may occur, particularly in dry weather and drought conditions, and accounts for the wilting (drooping) of plant tissues.

Osmosis plays a key role in three important processes in plants:

1. Water is taken up into root hairs from the soil and then travels to the root xylem by osmosis (page 136).
2. Water travels from the xylem into the mesophyll cells of the leaf by osmosis (page 137).
3. Stomata (page 49) open and close by a change in turgidity of their guard cells. Increased turgidity is achieved by increasing the osmotic potential of the cell contents so that water flows in by osmosis. When this happens the stomatal pore opens. When the O.P. of the guard cells is lowered, water is drawn out, the cells become flaccid, and the stomatal pore closes.

4.3. *Active Transport*

Active transport is a chemical process which results in a movement of particles in an opposite direction to that expected by diffusion. It occurs when molecules are taken across a membrane from a region of low concentration to a region of higher concentration, i.e. against a concentration gradient. Protein 'carrier' molecules within the cell membrane are thought to carry the substances across.

Active transport, as its name implies, is an active process and requires energy. This energy is obtained from respiration and any factors which locally influence the rate of respiration also influence the rate of active transport. Thus, low temperature, lack of oxygen and respiratory inhibitors such as cyanide, all inhibit active transport.

Two well-known examples of active transport are:

1. The active uptake of certain mineral salts by the root hair cells of a plant (see page 143).

2. The selective reabsorption of substances by mammalian kidney tubules (see page 155).

Read back over the chapter once again and then answer the questions below.

Definitions

Diffusion
Osmosis
Osmotic pressure

Key Words

Concentration (diffusion) gradient
Semi- (selectively) permeable membrane
Osmotic potential
Turgidity (turgor)
Plasmolysis
Active transport

Questions

1. (a) What do you understand by the term osmosis?
(b) How does diffusion differ from this?

2. In a single-celled organism living in water, explain how diffusion is involved in:

(*i*) *movement of oxygen*
(*ii*) *excretion of dissolved substances.*

(Look at page 317 if you need help in answering this question.)

3. (*a*) *Describe an experiment to demonstrate osmosis.*
(*b*) *What part does this process play in the uptake of water by a root-hair cell?*
(*c*) *Why are contractile vacuoles generally present in freshwater protozoa but absent from marine protozoa?*

(Look at pages 136 and 150 if you need help in answering parts (b) and (c) of this question.)

5. *Feeding (Nutrition)*

Nutrition is the study of food and feeding processes. Food is the material from which organisms obtain:

1. the energy
2. the raw materials

to construct, maintain and repair the body.

5.1. *Types of Nutrition*

There are two main types of nutrition: **autotrophic** (holophytic) nutrition, typical of green plants, and **heterotrophic** nutrition, typical of animals and non-green plants.

Autotrophic organisms make their own organic food from simple inorganic substances, using the energy from sunlight – a process called **photosynthesis**.

Heterotrophic organisms get their organic food 'ready-made'. Directly or indirectly, they are completely dependent on plants for it. All their organic food has originally been made by a green plant.

Heterotrophic nutrition can be divided into three types:

1. **Holozoic nutrition**: this is typical of most animals, ranging from *Amoeba* (page 315) to humans. It involves five stages:

(a) Complex organic food is taken into a cavity within the body (**ingestion**).

(b) Within the cavity, organic food is broken down into simple soluble substances (**digestion**).

(c) The soluble products of digestion are taken up into the organism (**absorption**).

(d) These soluble substances are incorporated into the protoplasm of cells (**assimilation**).

(e) Indigestible waste is removed from the body (**egestion**).

2. **Saprophytic nutrition**: this type of nutrition is shown by many non-green plants, including certain fungi, e.g. *Mucor* (page 301) and certain bacteria (page 294). The saprophytic organism (saprophyte) secretes

digestive enzymes on to dead or decaying organic matter. The organic food is digested and the soluble products of digestion are absorbed. Saprophytic nutrition differs from holozoic nutrition in that digestion occurs *outside* the organism and the organic food is always obtained from dead or decaying material.

Saprophytes play a very important role in the cycling of materials in nature (see page 269).

3. **Parasitic nutrition**: a parasitic organism (parasite) lives in or on another live organism (its **host**) from which it obtains its organic food. Examples of parasites are viruses (page 292), certain bacteria (page 294), and the fungus *Phytophthora* (page 302).

5.2. *Photosynthesis*

The process by which green plants make organic compounds from simple inorganic compounds is called **photosynthesis**. It can be represented by the following equation:

$$\text{carbon dioxide} + \text{water} \xrightarrow[\text{chlorophyll}]{\text{sunlight}} \text{glucose} + \text{oxygen}$$

$$6CO_2 + 6H_2O \rightarrow C_6H_{12}O_6 + 6O_2$$

This equation is really an oversimplification:

1. Glucose is not the only product of photosynthesis, though it is the main one. Other sugars are formed which can then be converted to fats or, by combining with mineral salts, form amino acids and vitamins. Photosynthesis is the source of all organic compounds in a plant.
2. Photosynthesis is not a one-step chemical reaction, but is really a series of over forty reactions, each reaction having a specific enzyme. All green parts of a healthy plant can carry out photosynthesis. The green coloration is due to the presence of the pigment **chlorophyll**, which traps sunlight energy. The chlorophyll converts this energy to chemical energy, which is used to bind atoms and small molecules together to make organic molecules.

In flowering plants (page 308) the leaves are the main sites of photosynthesis. Photosynthesis also occurs in the green stems of herbaceous plants (those flowering plants without woody stems, page 309).

The leaf of a herbaceous dicotyledon (page 308) is made up of a **midrib** and **lamina** attached to the stem by a stalk or **petiole**. A vein (vascular bundle) runs through the midrib and branches off into smaller veins in the lamina (see Fig. 5.1).

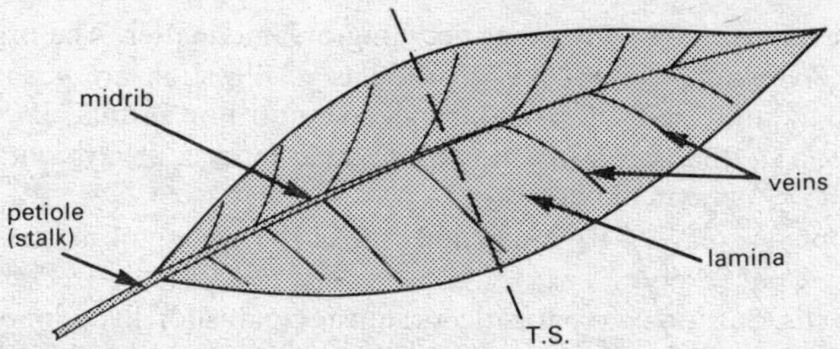

Fig. 5.1. *A leaf from a dicotyledon (see page 308)*

Within a cell of the leaf, **chloroplasts** are the actual sites of photosynthesis. It is these organelles (page 21) which contain chlorophyll together with the enzymes which catalyse the reactions of photosynthesis.

In transverse section (sometimes called a vertical section), the lamina of the leaf is seen to have four tissue layers covered on top and bottom by a thin waxy cuticle (see Fig. 5.2). Cells of the **palisade** layer contain the largest number of chloroplasts and it is here that the bulk of photosynthesis occurs. The **spongy mesophyll** cells contain fewer chloroplasts and the cells are separated by large air spaces which allow diffusion of gases through the leaf. The single layer of cells (**epidermis**) at the top and bottom of the lamina contains no chloroplasts except in the **guard cells** of the **stomata** (singular: stoma). Stomata are pores (holes) in the leaf surface which can be opened or closed by changes in turgidity of the guard cells which border them (see page 50). The pores allow carbon dioxide and oxygen to be exchanged between the leaf and the atmosphere. Stomata are usually found only on the lower surface of the leaf, away from the sun's direct rays. This reduces water loss by evaporation and diffusion through the stomatal pores.

Questions on leaf structure, and how leaf structure is related to photosynthesis, are popular with examiners. For example:

Q. Make a labelled diagram to show the arrangement of structures in the leaf of a flowering plant as seen in vertical section. Indicate on your diagram the distribution of chloroplasts.

Fig. 5.2 provides an effective answer.

Q. In what ways are the leaves of plants adapted for photosynthesis?

For photosynthesis to occur, carbon dioxide, water and light must come together at the site of photosynthesis. Let us examine how each of these factors reaches the chloroplasts in the leaf cells:

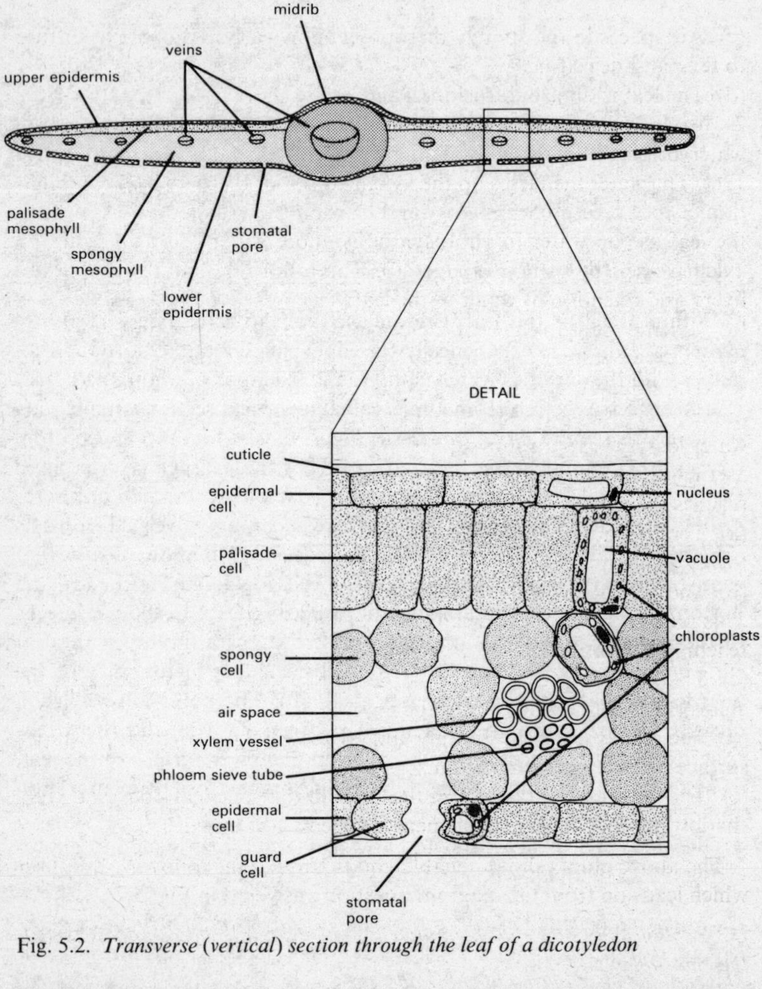

Fig. 5.2. *Transverse (vertical) section through the leaf of a dicotyledon*

1. **Carbon dioxide** diffuses from the atmosphere through the stomata and into the air spaces of the spongy layer. The gas dissolves in the moisture on the surface of mesophyll cells and diffuses to the chloroplasts. The following features of the leaf encourage this inward diffusion of carbon dioxide:

(a) The leaf is flat and has a large surface area.

(b) Stomata allow carbon dioxide to enter the leaf from the atmosphere.

(c) Air spaces in the spongy mesophyll allow carbon dioxide to diffuse to the palisade cells.
(d) The leaf is thin so diffusion distances are short.
2. **Water** is delivered to the leaf in the xylem vessels of the veins. The water originally enters the plant by osmosis through root hairs (page 136). The water then rises up the xylem vessels of the stem by a combination of root pressure, capillarity and transpiration (page 137). As cells in the leaf use up water in photosynthesis, more water is drawn from the xylem vessels by osmosis (page 40). The following features enable delivery and retention of water in the leaf:
(a) Xylem vessels in the veins deliver water to all parts of the leaf.
(b) The waxy cuticle covering the leaf reduces water loss by evaporation.
(c) The stomata, which allow gas exchange to occur, are found in larger numbers on the lower surface, away from direct sunlight. This reduces water loss by evaporation.
3. **Light**: the following features maximize the available light which reaches the chloroplasts in the palisade layer:
(a) The palisade cells are close to the upper surface of the leaf, directed towards the sunlight.
(b) The palisade cells are elongated towards the direction of sunlight. This reduces the number of cross walls light has to pass through before reaching chloroplasts.
(c) The large surface area of the lamina exposes a very large area of chlorophyll-containing cells to the light.
(d) The leaves are held at right angles to the light so that the upper surface receives maximum light.
(e) The leaves are arranged on the stem so that one leaf does not overshadow another. This arrangement is called a **leaf mosaic**.

The above points should enable you to answer the following question which leads on from the diagram question answered in Fig. 5.2:

Q. Explain fully how the raw materials required for photosynthesis reach the chloroplasts.

NOTE. Light is not a raw material.

5.3. *The Fate of the Products of Photosynthesis*

Glucose and other sugars

1. Much of the glucose is converted to starch for temporary storage in

the leaf. At night the starch is broken down to sucrose and transported round the plant in the phloem (page 145).
2. In the leaf and throughout the plant, sugars are broken down in respiration to release energy.
3. In growing regions, glucose is condensed to cellulose to make cell walls.
4. In the leaf, some sugars are combined with nitrates to form amino acids which are later incorporated in proteins.
5. In the leaf, some sugars are used to make fats and vitamins.

Remember, all cells in the plant will need a supply of organic substances. Photosynthesis is the original source of these. The substances are transported in the phloem to cells throughout the plant (see page 145).

Oxygen

1. Used in respiration, or
2. Excreted through the stomata.

5.4. *Photosynthesis Experiments*

In a healthy plant, an adequate supply of the following is required for photosynthesis to take place:

carbon dioxide
water
light
chlorophyll
plus, a suitable temperature.

These requirements can be demonstrated in a series of experiments. In such experiments, the presence of starch in a leaf is normally taken to indicate that photosynthesis has occurred. The assumption is that starch is a product of photosynthesis. This is generally the case: glucose formed in photosynthesis is converted to starch for temporary storage in the leaf.

Testing for starch

A green leaf is removed from a plant and the following steps are taken:
1. The leaf is placed in boiling water for five minutes to fix (kill) the leaf tissue and break down the waxy cuticle.

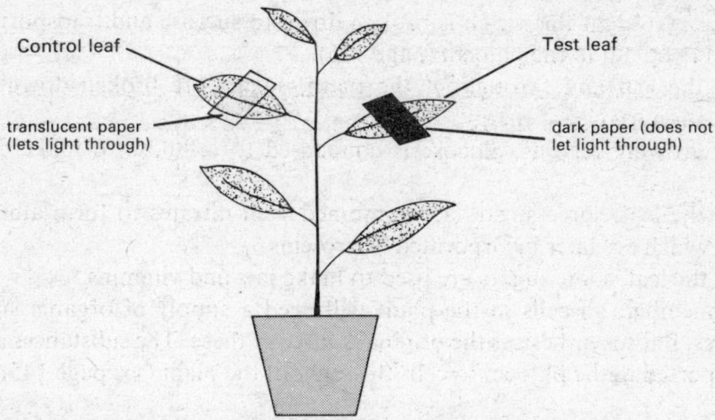

Fig. 5.3. *Apparatus testing the need for light in photosynthesis*

2. The leaf is transferred to boiling ethanol (alcohol) heated in a water bath to reduce the danger of fire. The ethanol removes chlorophyll. This decolourizes the leaf so that colour changes used in the starch test are not masked by the green colour of chlorophyll.
3. The decolourized leaf is put in hot water to soften it.
4. The softened leaf is spread out on a white tile.
5. It is covered for 1 minute with dilute iodine solution.
6. The iodine solution is then rinsed off with water.

A blue-black colouration indicates the presence of starch. If the brown colour of iodine persists, the test is negative for starch.

In experiments where the presence of starch is used to indicate photosynthesis, the leaves of the plant should be destarched before the experiment is begun. This is done by keeping the plant in darkness for 48 hours.

An experiment which demonstrates the need for light in photosynthesis

Method

1. Use a destarched plant.
2. Set up control and test leaves as shown in Fig. 5.3. For the test, the leaf is sandwiched between two sheets of dark paper, so that both top and bottom surfaces of the leaf are covered. For the control, transparent or translucent paper is used.
3. Give the plant water and leave in daylight or artificial light for six hours.

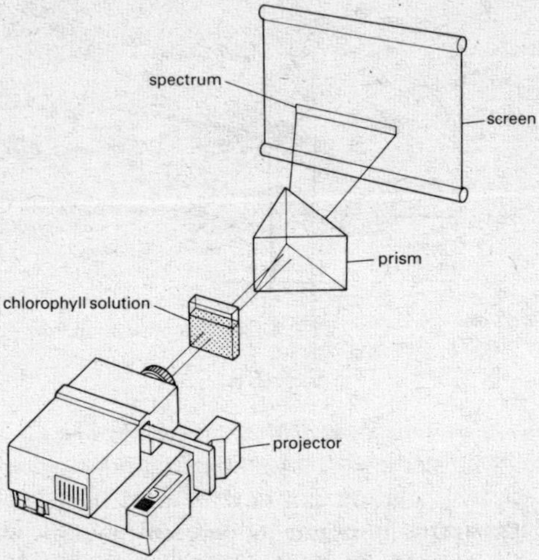

Fig. 5.4. *Demonstration of light absorption by chlorophyll*

4. Remove the test and control leaves and test them for starch (see starch test above).

Results

1. The covered strip on the control leaf turns blue-black with iodine (positive for the starch test).
2. The covered strip on the test leaf does not turn blue-black (negative for the starch test).

Conclusion

The absence of starch in the covered strip of the test leaf indicates that in the absence of light photosynthesis does not take place.

Light is needed for photosynthesis to occur.

An experiment which demonstrates the absorption of light by chlorophyll

A solution of chlorophyll is obtained in the following way:

1. Cut up a few leaves and place them in a mortar with a little sand.
2. Add ethanol to the mortar and grind with a pestle to break down the leaf cells and to dissolve the chlorophyll in the ethanol.
3. Filter the liquid from the mortar to separate the bits of leaf from the chlorophyll solution.
4. Pour the chlorophyll solution into a narrow container.

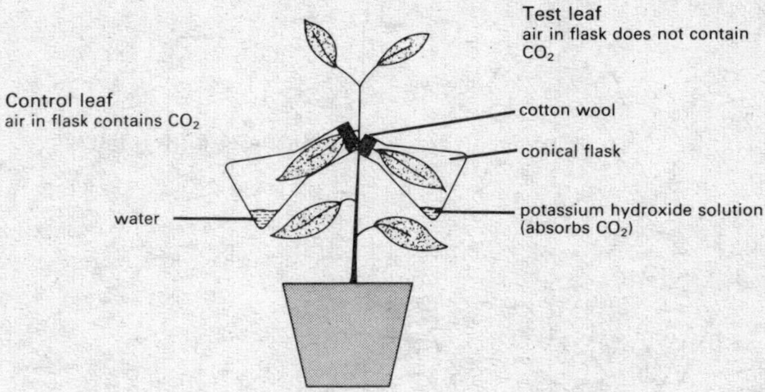

Fig. 5.5. *Apparatus testing the need for CO_2 in photosynthesis*

The chlorophyll solution can now be used in the following way. A prism will split white light into its different colours. By shining white light from a projector through the chlorophyll solution and then through a prism on to a screen, you can see which colours of light are absorbed by the chlorophyll. The apparatus is set up as shown in Fig. 5.4.

An experiment which demonstrates the need for carbon dioxide in photosynthesis

Method

1. Use a destarched plant.
2. Set up the control and test leaves as shown in Fig. 5.5.
3. Give the plant water and leave in daylight or artificial light for six hours.
4. Remove the test and control leaves and test them for starch (see starch test above).

Results

1. The control leaf turns blue-black with iodine (positive for the starch test).
2. The test leaf does not turn blue-black (negative for the starch test).

Conclusion

The absence of starch in the test leaf indicates that in the absence of carbon dioxide photosynthesis does not take place.

Carbon dioxide is needed for photosynthesis to occur.

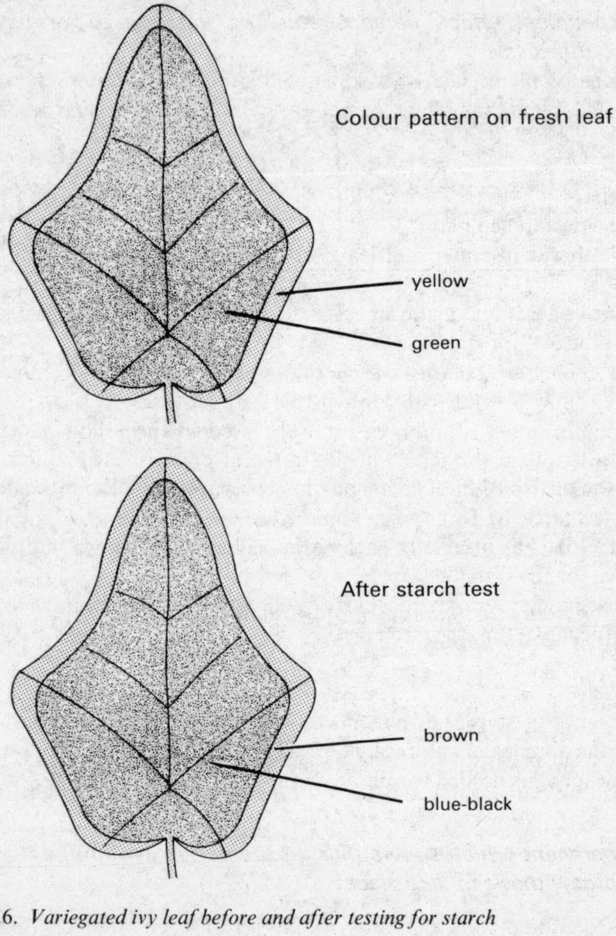

Fig. 5.6. *Variegated ivy leaf before and after testing for starch*

An experiment which demonstrates the need for water in photosynthesis

Such an experiment cannot easily be performed. The need for water in photosynthesis cannot be tested by depriving the leaf or plant of water. Without water, living tissue will die. In practice, sophisticated labelling experiments are required to show the need for water in photosynthesis.

An experiment which demonstrates the need for chlorophyll in photosynthesis

A variegated plant, such as a variegated strain of geranium, is used. A variegated plant has white or yellow chlorophyll-free patches on its leaves.

Method

1. Use a destarched plant.
2. Give the plant water and leave in daylight or artificial light for six hours.
3. Remove a leaf and make an accurate drawing of its colour patterning.
4. Test the leaf for starch.

NOTE. The green chlorophyll-containing part of the leaf is the control and the yellow or white chlorophyll-free part is the test.

Results

When the distribution of chlorophyll is compared with the results of the starch test (see Fig. 5.6), typical findings are:

1. Those patches originally green give a blue-black colour with iodine (positive for the starch test).
2. Those patches originally white or yellow do not turn blue-black with iodine (negative for the starch test).

Conclusion

The absence of starch in the chlorophyll-free part of the leaf indicates that in the absence of chlorophyll, photosynthesis does not take place.

Chlorophyll is needed for photosynthesis to occur.

An experiment which shows that a suitable temperature is required for photosynthesis to take place

In this experiment, rate of photosynthesis is measured. In theory, changes in the amounts of either the raw materials or products of photosynthesis could be used to measure the rate of reaction. In practice, it is easiest to measure the volume of gas (oxygen) given off by the plant in a given length of time. Oxygen is most easily collected over water, and so an aquatic plant is used.

Method

1. A sprig of Canadian pondweed, *Elodea*, is placed in the apparatus shown in Fig. 5.7.

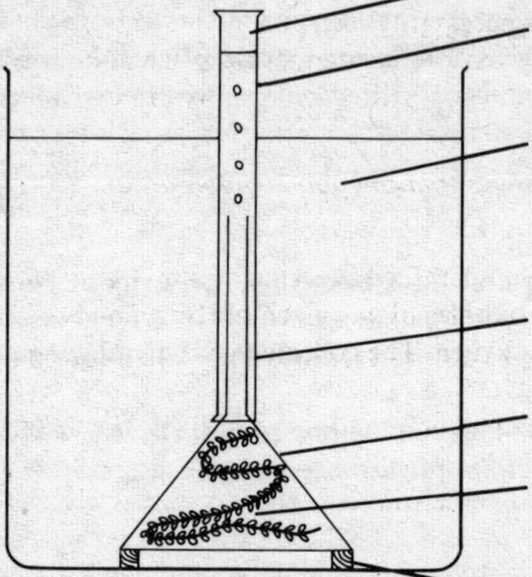

Fig. 5.7. *Apparatus for testing the effect of temperature on rate of photosynthesis*

2. The plant is exposed to a standard light source for a set time and the bubbles of gas given off during this time are collected in a measuring cylinder. The total volume of collected gas is recorded.
3. The experiment is conducted at different temperatures, starting at 15 °C and repeated at 10 °C intervals up to 45 °C. The apparatus is kept in a water bath to minimize temperature fluctuations.

Results

The rate of photosynthesis (as measured by volume of gas given off per hour or minute) increases with rise in temperature, up to 35 °C. At 45 °C very little gas is collected.

Conclusions

Rate of photosynthesis, as measured by gas evolved, is influenced by temperature. Between 15 °C and 35 °C, a rise in temperature increases the rate of photosynthesis. This is typical of most chemical reactions. When the temperature is raised, the reaction is speeded up. At 45 °C, photosynthesis is inhibited because the enzymes involved in photosynthesis are being denatured (see page 33).

NOTE. There are certain complications with this experiment:
1. Solubility of gases decreases with rise in temperature.

2. Gases increase in volume with rise in temperature.
3. Not all gas collected is necessarily oxygen.
4. Some of the oxygen given off in photosynthesis will be used up in respiration.

An experiment which shows that oxygen is given off in photosynthesis

The apparatus from the temperature experiment above can be used. In the presence of light, the bubbles of gas given off by a pondweed can be shown to contain largely oxygen. The collected gas will relight a glowing splint, a test for oxygen.

If the experiment is repeated in the absence of light, the collected gas will not relight a glowing split.

The two results taken together indicate that oxygen is given off by a photosynthesizing plant.

5.5. The Mechanism of Photosynthesis

NOTE. This is not required by all examining boards. Check the Contents list for your exam board requirements.

As mentioned earlier, the simple equation for photosynthesis gives no indication of the complexity of the reactions involved. Each reaction requires a specific enzyme, and all these enzymes are contained within chloroplasts. Photosynthesis occurs in two main stages, the light and dark reaction.

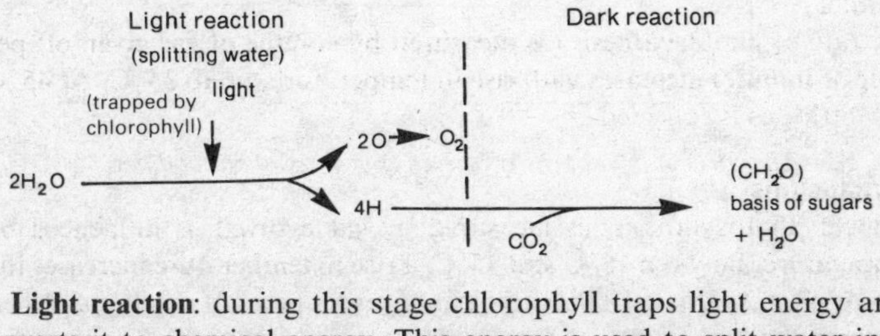

1. **Light reaction**: during this stage chlorophyll traps light energy and converts it to chemical energy. This energy is used to split water into oxygen and hydrogen. Oxygen is released as a waste product, and the hydrogen is used in the second phase of photosynthesis, the dark reaction.
2. **Dark reaction**: this stage does not require light, but depends on a supply of hydrogen from the light reaction. Hydrogen combines with

carbon dioxide to form sugars, which may afterwards be converted to other types of organic molecule.

5.6. *Limiting Factors and Photosynthesis*

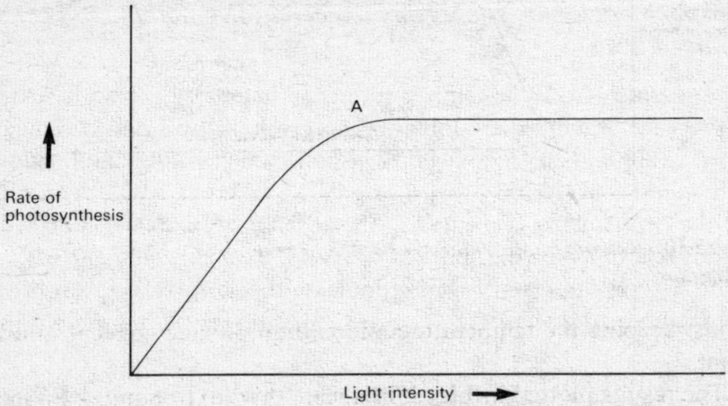

Fig. 5.8. *The effect of light intensity on rate of photosynthesis*

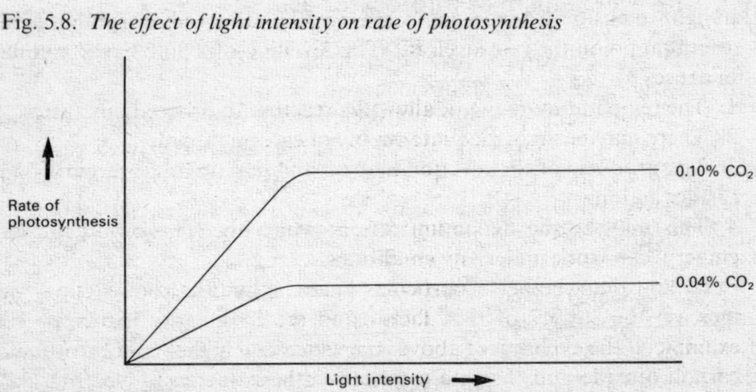

Fig. 5.9. *The effect of light intensity on rate of photosynthesis at different CO_2 concentrations*

In section 5.4 we examined the conditions which are necessary for photosynthesis to occur. If any one of these factors is in short supply, the rate of photosynthesis may be slowed down. The factor in shortest supply, and therefore the one controlling the rate of the reaction, is called the **limiting factor**.

The principle of limiting factors can be illustrated by placing a plant in increasing light intensities and measuring its rate of photosynthesis,

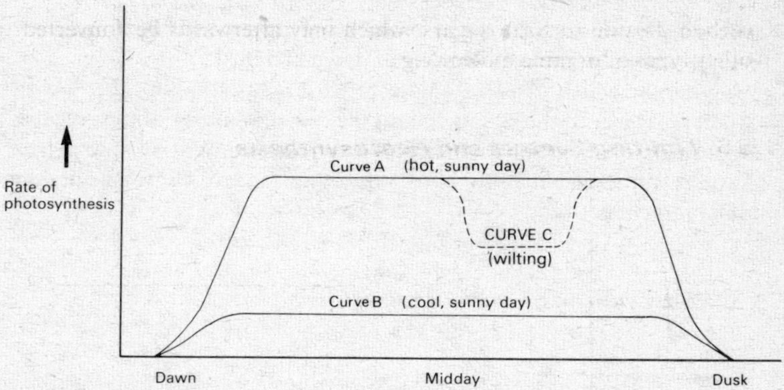

Fig. 5.10. *Rates of photosynthesis at different times of day, under different conditions*

whilst keeping the temperature and carbon dioxide concentration constant.

The results plotted in Fig. 5.8 indicate that up to point A an increase in light intensity produces an increase in the rate of photosynthesis, but after that point, the rate levels off. This levelling-off could have a number of causes:

1. The temperature may not allow the reaction to proceed any faster.
2. There may be an inadequate supply of carbon dioxide.
3. The products of the reaction may be building up and preventing any further action.
4. This may be the maximum rate at which the 'photosynthetic machinery' can work under any conditions.

To find out whether a particular factor is limiting the reaction, you increase the supply of that factor and see if the rate increases. For example, if the experiment above is repeated at a higher concentration of carbon dioxide and the rate of photosynthesis increases (see Fig. 5.9), then carbon dioxide was originally acting as a limiting factor.

If we consider a plant growing outdoors in a temperate climate, carbon dioxide concentration in the atmosphere varies little (a constant 0.04 per cent), whereas light and temperature vary with time of day and season of the year.

Curve A of Fig. 5.10 shows the rate of photosynthesis for a plant at different times on a hot, sunny day in summer. For some time after dawn, light and temperature are probably the limiting factors; during the middle hours of the day carbon dioxide is limiting; and at dusk, light is limiting.

On a cool sunny day (curve B) temperature may be limiting throughout the day, and light intensity limiting at dawn and dusk.

Any factor which causes stomata to close, thus stopping the flow of carbon dioxide into the leaf, will limit the rate of photosynthesis, regardless of the level of carbon dioxide in the atmosphere. Wilting, due to lack of water, will reduce the rate of photosynthesis (curve C) by causing the stomata to close.

5.7. *Photosynthesis and Respiration*

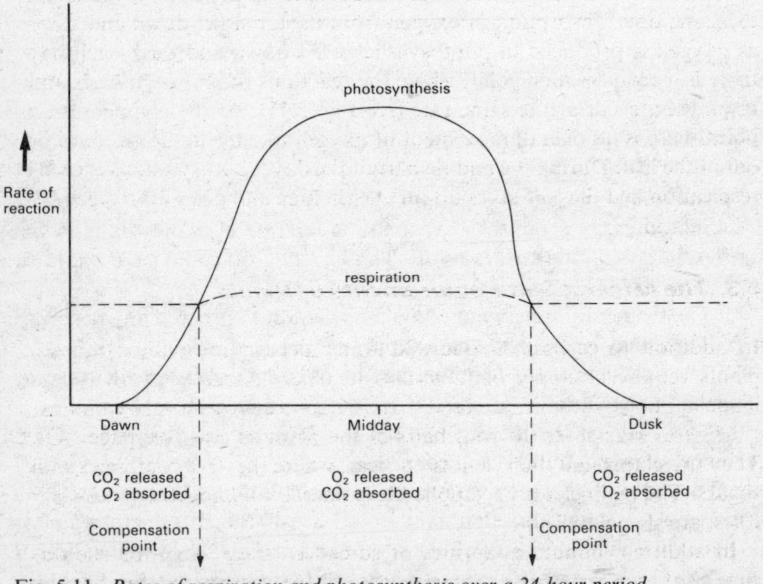

Fig. 5.11. *Rates of respiration and photosynthesis over a 24-hour period*

1. Photosynthesis is essentially the reverse of respiration. In photosynthesis, organic substances are built up and provide a store of chemical energy. In respiration, organic substances are broken down to release energy.
2. Photosynthesis uses the products of respiration while respiration uses the products of photosynthesis:

$$\text{carbon dioxide} + \text{water} + \text{energy} \underset{\text{respiration}}{\overset{\text{photosynthesis}}{\rightleftharpoons}} \text{glucose} + \text{oxygen}$$

3. Green plants (like all living things) respire all the time (both in the light and in the dark).
4. During daylight, plants photosynthesize as well as respire. For most of the daytime, photosynthesis proceeds at a much faster rate than respiration (see Fig. 5.11). More than enough food is made during the day to last through the night.

We can apply the above points to the events which occur in a leaf over a 24-hour period. In darkness, respiration alone occurs and the leaf takes in oxygen and releases carbon dioxide through the stomata. As dawn breaks, photosynthesis begins and the release of carbon dioxide slows down and stops as carbon dioxide is used up in photosynthesis. At the same time, absorption of oxygen from the air slows down and stops as oxygen is produced in photosynthesis. At dawn and dusk each day there is a **compensation point**, where the reactions of photosynthesis and respiration occur at the same rate (see Fig. 5.11). At the compensation point there is no overall movement of oxygen or carbon dioxide into or out of the leaf. During the middle part of the day, photosynthesis exceeds respiration and the leaf takes up carbon dioxide and gives off oxygen.

5.8. *The Mineral Salt Requirements of Plants*

In addition to carbon dioxide and water needed for photosynthesis, plants require a supply of mineral salts. These, together with sugars made in photosynthesis, are used to manufacture proteins and vitamins.

Mineral salts enter the root hairs of the plant as ions (see page 143). They travel through the xylem to the leaf, where they are combined with sugars. The main elements supplied by mineral salts, together with their roles, are shown in Table 5.1.

In addition, minute quantities of so-called 'trace' elements such as zinc (Zn), copper (Cu) and manganese (Mn) are needed for healthy growth. Large quantities are often poisonous.

Water culture experiments

The effect of a deficiency (lack) of a particular element on growth can be shown by 'water culture' experiments in which plants are grown in aerated solutions, rather than in soil.

For example, a batch of barley seedlings are grown in a series of solutions in glass bottles. A control plant is supplied with the complete

Table 5.1. *Elements required by plants and supplied in mineral salts*

Element		
Nitrogen (N)	As nitrate (NO_3^-), e.g. in KNO_3	In proteins and nucleic acids (DNA and RNA)
Sulphur (S)	As sulphate ($SO_4^=$), e.g. in $MgSO_4$	In proteins and nucleic acids (DNA and RNA)
Phosphorus (P)	As phosphate ($PO_4^{\equiv}$), e.g. in K_2PO_4	In enzymes, ATP and nucleic acids
Calcium (Ca)	As calcium ion (Ca^{++}), e.g. in $CaCO_3$	'Gum' (middle lamella) between adjacent cell walls
Iron (Fe)	As ferrous ion (Fe^{++}), e.g. in $FeSO_4$	In enzymes involved in making chlorophyll
Magnesium (Mg)	As magnesium ion (Mg^{++}), e.g. in $MgSO_4$	Part of chlorophyll molecule
Potassium (K)	As potassium ion (K^+), e.g. in K_2PO_4	Activates enzymes involved in respiration and photosynthesis

range of mineral salts. A series of test plants are grown in solutions in which one of the key elements is missing. In all other respects, the control and test plants are grown under the same set of conditions. After a few weeks the seedlings are examined and the growth of seedlings in test solutions compared with that of the control.

Typical results are as follows:

Experimental conditions	*Appearance of plant*		
	Stem	*Leaves*	*Roots*
Control (seedling grown in a solution containing N, S, P, Ca, Fe, Mg, K)	Tall, sturdy	Dark green	Normal growth
As control but no N	Thin and weak	Pale green or yellow	Normal
As control but no S	Thin and weak	Pale green or yellow	Normal
As control but no P	Short	Small and purple	Poor growth

As control but no Ca	No growth	–	Poor growth
As control but no Fe	Thin and weak	White	Normal
As control but no Mg	Normal	Yellow	Normal
As control but no K	Normal	Yellow with dead spots	Normal

In setting up water culture experiments certain precautions are necessary:
1. The water used to make up the stock solutions should, as far as possible, be pure.
2. The bottles and solutions should be sterilized to reduce the likelihood of fungal or bacterial growth.
3. The bottles should be wrapped in dark paper to prevent algal growth.

5.9. *Holozoic Nutrition*

All animals feed on ready-made organic food supplied by green plants. Animals cannot photosynthesize and are unable to make their own organic materials from simple inorganic molecules.

Carnivores (e.g. the dog, page 87) feed mainly on animals.

Herbivores (e.g. the sheep, page 87) feed on plants.

Omnivores (e.g. humans, page 67) feed on plants and animals.

Before food can properly enter the body it must pass across a cell membrane. In animals, complex organic food can enter cells of the body only if first broken down into small, soluble molecules. The stages involved in this process are as follows:
1. **Ingestion**: the intake of complex organic foods into a cavity within the body.
2. **Digestion**: the breakdown of complex organic foods into small, soluble molecules.
3. **Absorption**: the uptake of soluble food substances into the body across a cell membrane.
4. **Assimilation**: the uptake and use of soluble food substances by cells in the body.
5. **Egestion**: the removal of undigested food from the body. Do not confuse egestion with excretion (page 149) or secretion (below).

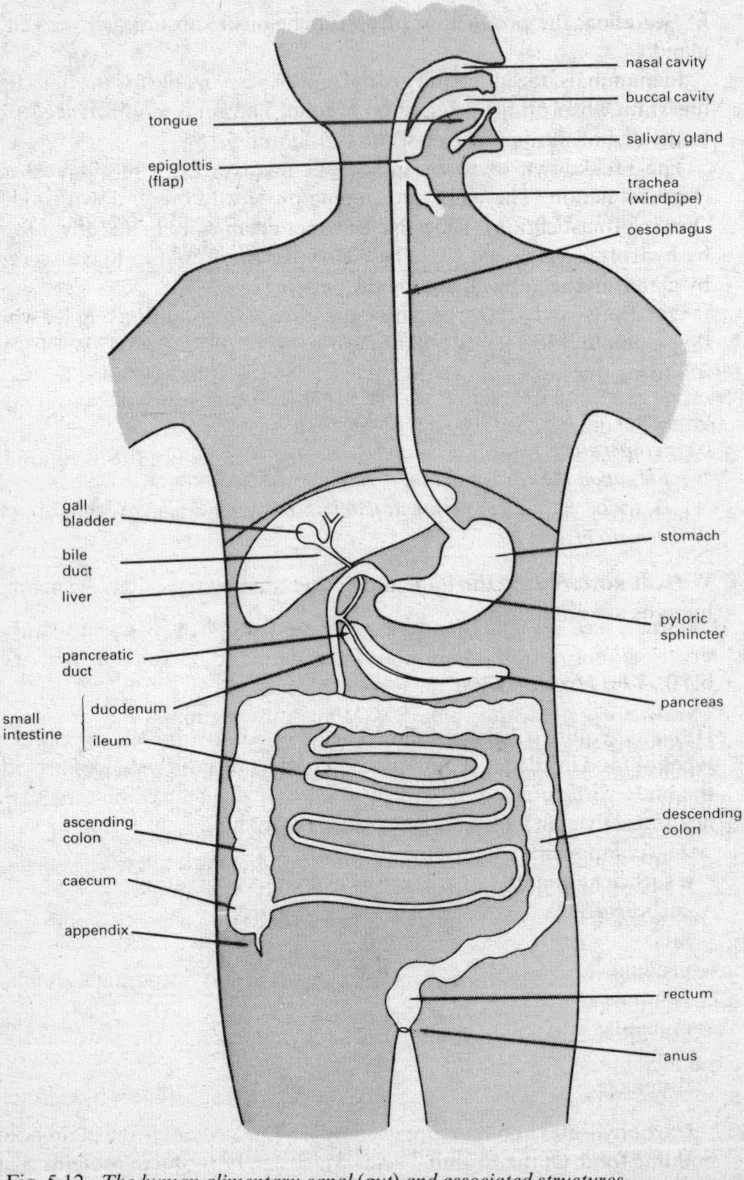

Fig. 5.12. *The human alimentary canal (gut) and associated structures*

6. **Secretion**: the production of useful chemical substances by a cell or gland.

In mammals, the alimentary canal (gut) is responsible for the ingestion, digestion, absorption and egestion of food. The gut is modified according to the diet of the animal (see sections 5.11 and 5.15).

The breakdown of foods in the gut involves both mechanical and chemical action. The teeth and tongue physically break down food by chewing (mastication). Digestive enzymes chemically break down foods by hydrolysis (see page 28). These enzymes are found in juices secreted by glands in the gut wall and by the pancreas.

Questions on digestive enzymes, together with conditions under which they operate, are very popular. Turn back to page 33 and then try the following question, using a digestive enzyme as your example.

Q. (a) What are the effects of a change in
(i) temperature
(ii) pH, upon the rate of action of any named *enzyme.*
(b) Describe with full experimental details how you would investigate either of these effects.

We will now look at the human diet and alimentary canal. Remember, humans are omnivores.

5.10. *The Human Diet*

Humans require a **balanced diet**. This is one that supplies the different types of food in adequate amounts and correct proportions, and provides the body with sufficient energy for its needs. A balanced diet maintains healthy active life and, where necessary, growth.

The components of a balanced diet are:
a sufficient quantity of energy (measured in kilojoules)
carbohydrates
fats
proteins
vitamins
minerals
water
roughage.

Carbohydrates and fats supply energy, while protein is the main body-building food. Of the 20 amino acids which can be found in proteins, 8 are essential in the human diet. From these 8 any of the other 12 can be made.

Table 5.2. *The different parts of the human diet*

Type of food	*Good sources*	*Main uses in the body*
Carbohydrates:		
Starch	Potatoes, flour products	As an energy source Stored as glycogen
Disaccharides	Table sugar, milk	
Monosaccharides	Honey, fruit	
Proteins:		
First-class proteins (rich in essential amino acids)	Meat, eggs, milk, cheese	Structural material in cells. Enzymes. General growth and repair of tissues
Second-class proteins	Peas, beans	
Fats (lipids)	Unskimmed dairy products; e.g. butter. Plant oils	As an energy source. As structural material in cell membranes. Stored as fat in adipose tissue
Vitamins At least twenty vitamins are required in minute quantities. Some are listed in Table 5.3.		
Minerals About fifteen different minerals are required in very small quantities. The major ones are listed in Table 5.4.		
Water	Liquids and most foods	Body 60–80 per cent water. All chemical reactions take place in it
Roughage	Cereals, vegetables	Indigestible. No food value. Stimulates muscular movements (peristalsis) of gut wall. Lack of roughage can lead to constipation

Proteins rich in essential amino acids are called first-class proteins.

Fat contains twice as much energy per gram as carbohydrates and proteins. This is why it is such a good store of energy.

The energy content of foods is measured in kilojoules per gram (kJ/gm).

4.2 joules of heat energy are required to raise the temperature of 1 gram of water by 1 °C.

1,000 joules = 1 kilojoule.

The amount of energy required in the diet varies with sex, age and physical activity.

Table 5.3. *Vitamins in the human diet*

Vitamin	*Good sources*	*Main uses in the body*
A (fat soluble)	Butter, milk, certain vegetables, e.g. carrots	Maintaining health of mucous membranes. Lack of vitamin (deficiency) increases likelihood of disease infection. Forms part of visual pigment in rod cells of eye (page 175). Lack of vitamin leads to night blindness
B (water soluble) There are over ten B vitamins, two of which are thiamin (B_1) and niacin	Whole grain of cereal, such as in wholemeal bread. Liver, yeast	Thiamin and niacin play a role in energy release (respiration). Lack of thiamin leads to paralysis (**beri-beri**). Lack of niacin leads to gut disorders and the skin disease **pellagra**
C (water soluble) Destroyed by boiling or prolonged exposure to sunlight	Fresh citrus fruits. Certain fresh vegetables, e.g. potatoes. Milk	Repair of tissue damage. Maintaining health of skin and blood vessels. Lack of vitamin leads to **scurvy** (bleeding gums, poor wound healing, and bleeding under skin)
D (fat soluble)	Liver, fish-liver oil, dairy products. Also manufactured by the skin when exposed to u.v. light	Used in bone formation. Lack of vitamin leads to skeletal deformity such as **rickets** in children

	Energy requirement (average)
Adult male office worker	12,000 kJ/day
Adult female office worker	9,500 kJ/day
Adult male manual worker	15,000–20,000 kJ/day
Male 15 years old	11,000 kJ/day
Male 10 years old	9,000 kJ/day
Male 5 years old	6,500 kJ/day

Table 5.4. *Minerals in the human diet*

Mineral	*Good sources*	*Main uses in the body*
Calcium (Ca)	Dairy products, bread, fruit and vegetables	Bone and teeth formation. Lack of mineral leads to brittle bones and teeth
Iodine (I)	Table salt. Sea foods	In thyroid gland hormone, thyroxine. Lack of mineral leads to thyroid gland swelling (goitre)
Iron (Fe)	Liver, egg yolk	Formation of haemoglobin (see page 117)
Phosphorus (P)	Dairy products	Bone and teeth formation. Lack of mineral leads to brittle bones and teeth
Potassium (K) Sodium (Na)	Plant material Table salt	Important constituents of blood and cells. A proper balance of these is required for proper functioning of nerves and muscles

Too much energy-giving food will cause overweight, while too little will cause underweight.

The detailed nutritional requirements for a human are shown in Tables 5.2, 5.3 and 5.4. Short-answer questions on these are common.

Questions on food tests are also popular. Refer back to page 31 and then answer this question.

Q. Describe, for each of the following, one *test you could carry out in order to demonstrate their presence in food materials:*

(*i*) *reducing sugar, e.g. glucose*

(*ii*) *fats*

(*iii*) *proteins.*

Malnutrition

Malnutrition is a result of not having a properly balanced diet. If the body does not receive the correct nutrients in the right proportions it cannot function properly.

Many illnesses are caused by not eating certain vitamins and minerals (see Tables 5.3 and 5.4). In the tropics, and sub-tropics, diets are often lacking in protein, and a resulting disease is called **kwashiorkor**. This disease occurs in young children when they stop breast feeding and are weaned on to food that does not contain enough protein for their growth requirements.

Obesity, or being overweight, is a common form of malnutrition in Western countries. It occurs when the amount of food eaten is more than the body needs to make energy. The excess food is changed to fat and stored in special tissues under the skin called **adipose tissue**. Obesity can cause many health problems and is a major cause of high blood pressure (hypertension) in the West.

The problems of food supply in developing countries

1. **Weather**. The weather in many developing countries is unpredictable. Water is necessary for plant growth, therefore countries which experience drought cannot rely on growing enough food each year.
2. **Population growth**. Many developing countries have a rising population, so their food requirements increase each year.
3. **Overfarming**. To increase productivity farmers often use land that is not suitable for crops and the soil is quickly deprived of nutrients. This problem could be solved by the use of fertilizers but they are expensive.

   Overfarming may also result in erosion. Many forests have been cleared to make way for agricultural land. The root systems of food crops are not as extensive as those of trees so the soil is not held together as well and can be blown away by the wind.
4. **Inappropriate strains of crops**. Many strains of crops used in developing countries have been imported from other countries. Very often these strains are not suited to their new environment.
5. **Inappropriate technology**. Many farmers in developing countries have been provided with farm machinery by aid organizations. Very often, spare parts are not available for these machines when they break down. In many situations it is more efficient to use livestock to pull a plough.
6. **Getting used to aid**. Farmers who regularly receive food from aid organizations are less inclined to grow their own food. Aid is now aimed more at helping farmers to improve their efficiency and become self-sufficient.

Fig. 5.13. *Peristalsis in the oesophagus*

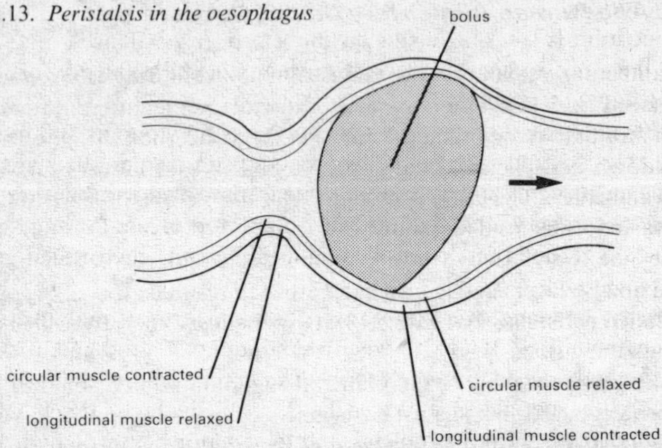

7. **Cultural barriers**. Even when food is available there may be malnutrition among certain peoples because of traditional eating practices. For example, Hindus are vegetarian and need to eat a wide variety of plants to obtain the full range of amino acids. If the food supply is limited to certain crops they may not receive a properly balanced diet. In some tribes in South Africa, women are not allowed to eat fowl or eggs. In Tanzania, some women are not allowed to drink milk.

Some solutions to these problems

1. **Irrigation**. One way to alleviate the problems of drought is to install irrigation systems to carry water from rivers or wells to the land.
2. **Research**. There are several ways in which research can help to improve the yields of food obtained by farmers.

Development of new strains of food plants can help these crops survive adverse conditions and therefore produce more food.

Research into pesticides is producing safer chemicals which can reduce pest populations without harming the farmers.

Development of appropriate machinery *and* training people to service such machines could improve the efficiency of farming in certain areas. Cheaper drilling methods can also provide access to more wells.

3. **Better storage**. When conditions are good there may be a surplus of food. Storing this food could help people survive a bad year later. There is, therefore, a need to develop cheap and effective methods of storage.

5.11. *The Human Alimentary Canal (Gut)*

The alimentary canal (gut) is a muscular tube running from mouth to anus (see Fig. 5.12). Different parts of the tube serve different functions. From mouth to duodenum, the gut is mainly concerned with the physical breakdown and chemical digestion of food. Absorption occurs in the ileum and colon. The last section of the gut, the rectum, prepares indigestible waste for disposal (egestion).

The gut wall has two layers of muscle, circular and longitudinal. These muscle layers act against each other to produce waves of contraction called **peristalsis** (see Fig. 5.13). By peristalsis, food is pushed along through the gut.

Mucus is secreted by the gut lining throughout its length and lubricates and protects the lining against abrasion.

The functions of the different parts of the gut can be summarized as follows:

Table 5.5. *The functions of the different regions of the human alimentary canal (gut)*

Region	*Functions*
Mouth (buccal cavity)	Ingestion, mastication (chewing), bolus (food ball) formation, some starch digestion
Oesophagus	Delivery of food from mouth to stomach
Stomach	Storage and delivery of food to small intestine. Formation of chyme (liquefied food). Some protein digestion. Acid kills bacteria
Duodenum	Digestion of starch, protein and fat
Associated organs: Liver	Production of bile which emulsifies (breaks up) fat
Pancreas	Production of pancreatic juice which contains enzymes for digesting starch, protein and fat
Ileum	Completes digestion of carbohydrates, proteins and fats. Food absorption occurs here
Caecum	No particular function in humans
Appendix	No particular function in humans
Colon	Water absorption
Rectum	Formation and storage of faeces (solid waste)
Anus	Egestion

Questions on events in different parts of the gut are fairly common. For example:

Q. Describe four important processes concerned with digestion that take place in the buccal cavity.

Mouth (buccal cavity)

Food is **ingested** through the mouth, where the teeth play a large part in preparing the food for digestion. **Mastication** (chewing) is the mechanical breakdown of food by the teeth, aided by the tongue and jaw muscles. The food is broken into small pieces thus exposing a larger surface area for the action of digestive enzymes.

Saliva, secreted by salivary glands
1. lubricates the food for swallowing
2. contains the enzyme salivary amylase (ptyalin) which digests starch to maltose
3. dissolves soluble food.

The teeth and tongue mould the food into a ball (bolus) which is pushed backwards against the roof of the mouth. This sets off the swallowing reflex whereby the bolus is pushed into the oesophagus by the contraction of throat muscles. At the same time, the soft palate closes over the nasal passages and the epiglottis closes over the entrance of the windpipe (trachea) thus stopping food 'going down the wrong way'.

Oesophagus

Inside the oesophagus, the bolus is squeezed down to the stomach by peristalsis (see Fig. 5.13).

Stomach

The stomach is a muscular bag. Its main functions are:
1. To store food temporarily and deliver it, a little at a time, to the duodenum. This prevents the digestive system becoming overloaded with too much food at once.
2. Muscular contractions of the stomach wall churn the solid food into a liquid called **chyme**.
3. The digestion of protein begins here.

The lining of the stomach contains gastric glands which secrete a number of substances:

1. **Hydrochloric acid**, which provides the correct pH for pepsin action (see 2), stops activity of the enzyme salivary amylase, and kills bacteria.
2. The enzyme **pepsin** which digests proteins to polypeptides and peptides.
3. **Rennin**, which is valuable to infants living on milk. It clots milk, so keeping it in the stomach where pepsin can break it down.

At intervals, a ring of muscle, the pyloric sphincter, at the lower end of the stomach relaxes, allowing small quantities of chyme through to the duodenum.

Duodenum

The duodenum is the first 30 cm (12 inches) of the small intestine.

Bile is a green fluid produced by the liver and stored in the gall bladder. It empties into the duodenum along the bile duct (see Fig. 5.12). Bile contains sodium bicarbonate, which neutralizes stomach acid, and bile salts, which emulsify fats (breaks them down into small droplets). Bile does *not* contain any enzymes.

Pancreatic juice, produced by the pancreas, enters the duodenum along the pancreatic duct. It contains sodium bicarbonate to neutralize stomach acidity, as well as three enzymes which together digest the three main types of organic food – carbohydrates, proteins and fats:

$$\text{Starch} \xrightarrow{\text{pancreatic amylase}} \text{maltose}$$

$$\text{Polypeptides} \xrightarrow{\text{trypsin}} \text{peptides} + \text{amino acids}$$

$$\text{Fats} \xrightarrow{\text{lipase}} \text{fatty acids} + \text{glycerol}$$

Ileum

The second part of the small intestine, the ileum, is where digestion is completed and absorption occurs. Here, the pancreatic enzymes are still active. In addition the glands in the ileum lining secrete enzymes which digest the three main types of organic food down to their soluble constituents:

$$\text{Sucrose} \xrightarrow{\text{sucrase}} \text{glucose} + \text{fructose}$$

$$\text{Maltose} \xrightarrow{\text{maltase}} \text{glucose}$$

$$\underset{\text{(disaccharides)}}{\text{Lactose}} \xrightarrow{\text{lactose}} \underset{\text{(monosaccharides)}}{\text{glucose + galactose}}$$

$$\text{Peptides} \xrightarrow[\text{of enzymes)}]{\text{erepsin (a mixture}} \text{amino acids}$$

$$\text{Fats} \xrightarrow{\text{lipase}} \text{fatty acids + glycerol}$$

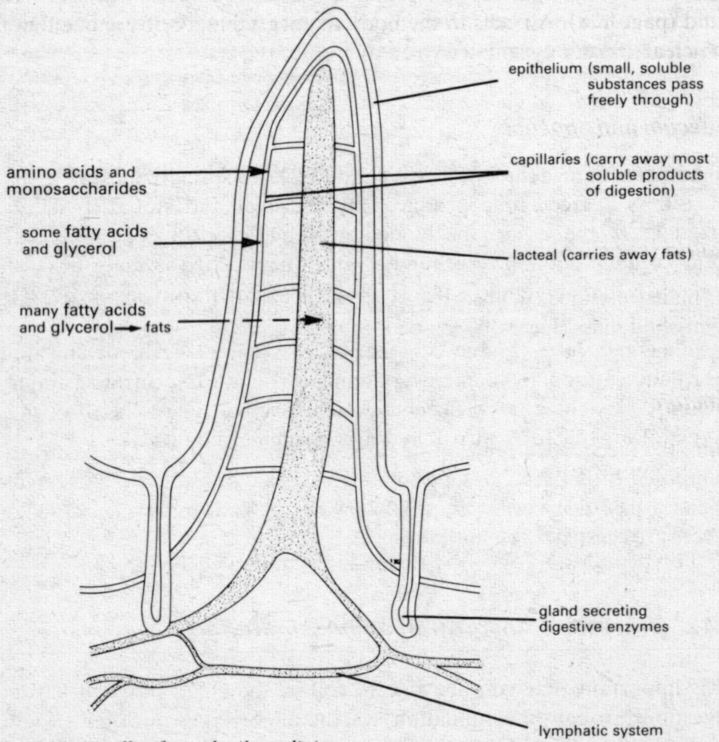

Fig. 5.14. *A villus from the ileum lining*

The organic molecules which were originally large and complex are now small and soluble and can be absorbed.

The lining of the ileum is folded into finger-like projections called **villi** (singular: villus). These greatly increase the surface area for absorption. Other features of the villi which aid absorption are shown in Fig. 5.14.

The final products of digestion are taken into the bloodstream across the lining of the villi. Absorption is by a combination of passive diffusion and active uptake (see page 44). Some fatty acids and glycerol recombine in the intestinal epithelium and as fats enter the lacteals in the centre of the villi (see Fig. 5.14). They then travel through the lymphatic system (page 129) before entering the bloodstream. All other products of digestion travel in the bloodstream to the liver (page 81) where their rate of supply to body tissues is regulated. The digested food is distributed round the body in the bloodstream and is delivered to cells via the tissue fluid (page 128). All cells in the body require a supply of organic foods which are finally **assimilated** when they enter the cell.

Caecum and appendix

In humans, these are reduced in size and have no special function.

Colon (large intestine)

Its main function is to absorb water from the fluid contents. The resulting semi-solid mass (**faeces**) is passed into the rectum by peristalsis.

Rectum

Here the faeces are stored before removal from the body. The faeces are composed of indigestible food (mainly cellulose and other plant fibres), bacteria and dead cells scraped from the gut lining. At intervals, the faeces are **egested** through the anus by muscular action.

5.12. *Digestion, Absorption and Assimilation*

It is important that you are able to follow the chain of events, from ingestion through to assimilation, for the major types of organic food. For example:

Q. Describe the parts played by the following in the digestion and absorption of protein in the diet of a mammal:
(i) the stomach
(ii) the pancreas
(iii) the small intestine.

Q. Name and give the action of the various enzymes in the digestion of carbohydrates in a mammal.

Carbohydrates

Carbohydrates are ingested as monosaccharides, disaccharides or polysaccharides. The last two need to be digested to monosaccharides before absorption.

Digestion

In the mouth (buccal cavity):

$$\text{starch} \xrightarrow[\text{(ptyalin)}]{\text{enzyme salivary amylase}} \text{maltose}$$

In the stomach: no carbohydrate digestion.

In the duodenum:

$$\text{starch} \xrightarrow{\text{enzyme pancreatic amylase}} \text{maltose}$$

In the ileum:

$$\text{maltose} \xrightarrow{\text{maltase}} \text{glucose}$$

$$\text{sucrose} \xrightarrow{\text{sucrase}} \text{glucose + fructose}$$

$$\underset{\text{(disaccharides)}}{\text{lactose}} \xrightarrow[\text{enzymes secreted by lining of ileum}]{\text{lactase}} \underset{\text{(monosaccharides)}}{\text{glucose + galactose}}$$

All digestible carbohydrates are now monosaccharides and can be absorbed across the lining of the ileum.

Absorption

In the ileum, monosaccharides (mainly glucose) pass through the epithelium of villi, through capillary walls and into the blood plasma. Absorption is by a combination of passive diffusion and active uptake. The glucose then travels in the hepatic portal vein to the liver.

Assimilation

What happens to glucose on reaching the liver is determined by the blood glucose level.

1. Glucose is the body's main source of energy. All cells need a supply of glucose for respiration (see page 92).
2. Excess glucose is converted to glycogen in the liver and skeletal muscles. When the blood glucose level drops, the liver converts its

glycogen back into glucose and releases it into the bloodstream. In this way, the blood glucose level is kept relatively constant.
3. Glucose in large excess is converted to fat and stored in adipose tissue beneath the skin (page 157) and around certain internal organs. Hence, a large amount of carbohydrate in the diet may lead to obesity. The fat may be respired later (see page 81).

Proteins

Proteins are ingested in the form of polypeptide chains of varying length. These have to be broken down to amino acids before absorption.

Digestion

In the mouth: no protein digestion
In the stomach:

$$\text{proteins} \xrightarrow{\text{enzyme pepsin}} \text{polypeptides}$$

$$\text{milk protein} \xrightarrow{\text{rennin}*} \text{clotted protein}$$

In the duodenum:

trypsinogen (inactive enzyme)
↓
trypsin (active enzyme)
↓

$$\text{proteins} + \text{polypeptides} \xrightarrow[\text{(in pancreatic juice)}]{} \text{peptides} + \text{amino acids}$$

In the ileum (trypsin still active):

$$\text{peptides} \xrightarrow[\text{secreted by the lining of the ileum)}]{\text{erepsin (several enzymes}} \text{amino acids}$$

All proteins have now been digested to amino acids which can be absorbed across the lining of the ileum.

Absorption

As with monosaccharides, amino acids pass into capillaries in the ileum wall by a combination of passive and active processes. The amino acids then travel in the hepatic portal vein to the liver.

* In children only. Milk is clotted so that it remains in the stomach where it can be digested by pepsin.

Assimilation

What happens to amino acids on reaching the liver depends on the body's requirement for protein.

1. All cells in the body need a supply of amino acids to build up their own proteins.
2. Excess amino acids are *not* stored, but in the liver are **deaminated** (their amine group removed). The amine group (NH_2) is altered to the very poisonous ammonia (NH_3) which is then converted to the less harmful substance **urea**. Urea is released into the bloodstream, transported in the blood to the kidneys, and then excreted in the urine (see page 155).

The remainder of the amino acid molecule is broken down in the liver to release energy in respiration.

Fats

Fats, unlike carbohydrates or proteins, do not readily mix with water but tend to stay clumped together as globules. Before fats can be effectively digested they must be physically broken down into smaller droplets (emulsification). This increases their surface area exposed to digestive enzymes.

Digestion

In the mouth: no fat digestion.
In the stomach: no fat digestion.
In the duodenum:

$$\text{fat globules} \xrightarrow{\text{bile salts}} \text{fat droplets}$$

$$\text{fat} \xrightarrow[\text{pancreatic juice}]{\text{lipase in}} \text{fatty acids + glycerol}$$

In the ileum (pancreatic lipase still active):

$$\text{fats} \xrightarrow[\text{lining of ileum}]{\text{lipase secreted by}} \text{fatty acids + glycerol}$$

Absorption

Only some of the fatty acids and glycerol enter the capillaries of the villi inside the ileum. A large proportion join together again in the intestinal epithelium and pass as fats into the **lacteals**. The fluid in the lacteals enters the lymphatic system which eventually empties its contents into the bloodstream (page 129). Thus, most of the products of fat digestion bypass the liver.

Assimilation

1. All cells in the body need fats to construct and repair membranes in the cell.
2. Fats are an important source of energy released by respiration.
3. Fats not immediately broken down in respiration are stored in adipose tissue under the skin and around certain internal organs. When required for energy, the stored fats are sent to the liver where they are converted to substances which can be readily respired.

5.13. ***The Liver***

The liver is the largest organ in the body. It has over 500 functions. It is reddish-brown and lies at the top of the abdomen just below the diaphragm.

The liver has the following connections with other parts of the body:

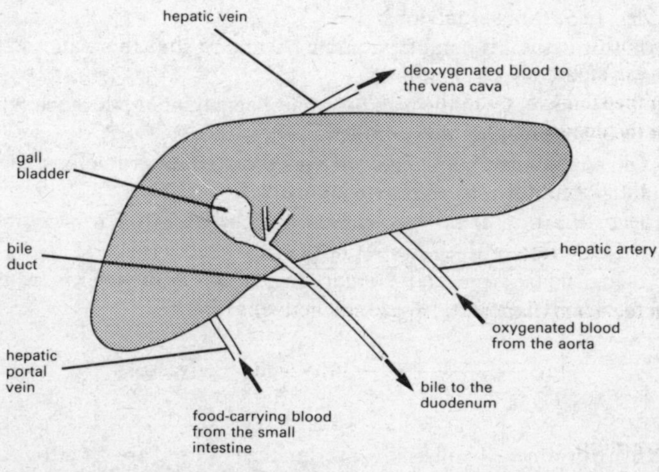

Fig. 5.15. *Connections to and from the liver*

The **hepatic artery** brings oxygenated blood from the aorta while the **hepatic portal vein** brings absorbed food from the alimentary canal.

All blood leaving the liver travels along the **hepatic vein** to the vena cava. Bile, which is stored in the gall bladder, empties into the duodenum along the **bile duct**.

The liver is a major homeostatic organ, i.e. it helps keep conditions within the body constant. Dissolved food substances absorbed through the gut wall are delivered in the blood straight to the liver. The liver then controls the 'through-flow' of these substances according to the body's needs.

Functions of the liver

The ten main functions of the liver are more easily remembered if grouped into categories.

Three functions concern the major types of organic foods – carbohydrates, proteins and fats:

1. The liver controls blood glucose level (see page 78).
2. It deaminates excess amino acids (see page 80).
3. It controls the body's use of fats (see page 81).

It has two manufacturing (synthetic) functions:

4. It forms plasma proteins, e.g. fibrinogen (see page 119).
5. It forms bile (see page 75).

It has two storage functions:

6. It stores the fat soluble vitamins A and D and the water soluble vitamin B_{12}.
7. It stores iron from the breakdown of haemoglobin (see 8 below).

It has two 'breaking-down' functions:

8. Old red blood cells are removed from the circulation and their haemoglobin broken down by the liver.
9. Detoxification. It converts poisonous substances to harmless ones.

One heat-related function:

10. Being such a large, active organ, respiration in the liver is the main source of body heat.

5.14. *Teeth*

Teeth are produced by the skin covering the jaws and are set into sockets in the jawbone.

The number and arrangement of teeth in the jaws is called the **dentition**.

The teeth of mammals differ from those of other vertebrates:

Mammals	*Other vertebrates*
Teeth have a **pulp cavity**	No pulp cavity
Teeth specialized for different functions (**heterodont** dentition)	Teeth all of the same type (**homodont** dentition)

In mammals, a set of **milk teeth** develop first and are later pushed out and replaced by **adult (permanent) teeth**.

Mammalian teeth are layers of modified bone tissue, supplied with food and oxygen by the pulp cavity (see Fig. 5.16), and shaped according to function.

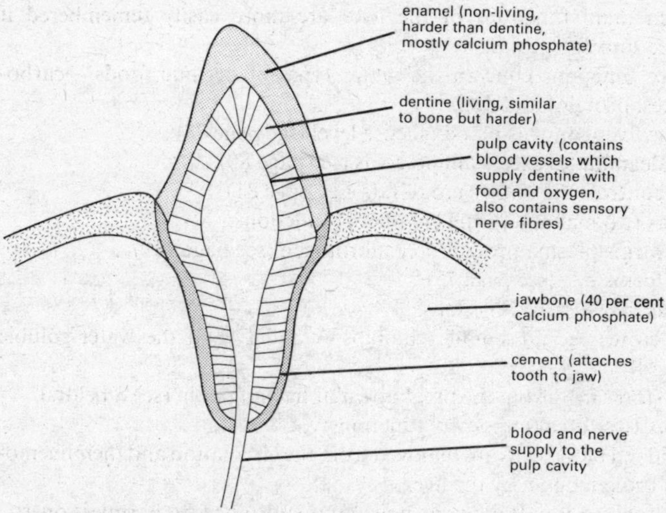

Fig. 5.16. *V.S. of canine tooth of a mammal*

There are four types of teeth in mammals:

Type	*Position in the jaw*	*Appearance*	*Function*
Incisor	At the front	Chisel-shaped, single root	Biting or gnawing
Canine	Behind the incisors	Pointed, single root	Stabbing or tearing
Premolar (cheek teeth)	Behind the canines	Flattened with 2 cusps, 1 or 2 roots	Grinding, crushing or slicing
Molar (cheek teeth)	Behind the premolars	Flattened with 4 cusps, 2 or 3 roots	,, ,,

The number and type of teeth are determined by the mammal's diet.

1. **Humans** are **omnivores** and have a 'generalized dentition':

(a) The incisors are flat for biting all types of food.

(b) Canine are present but are not large.

(c) The cheek teeth have rounded cusps for chewing a mixed diet.

2. A **dog** is a **carnivore**. Its dentition is adapted for seizing and killing prey, tearing and chewing flesh, and cracking and crushing bones. Refer to Fig. 5.17 as you go through the following points:

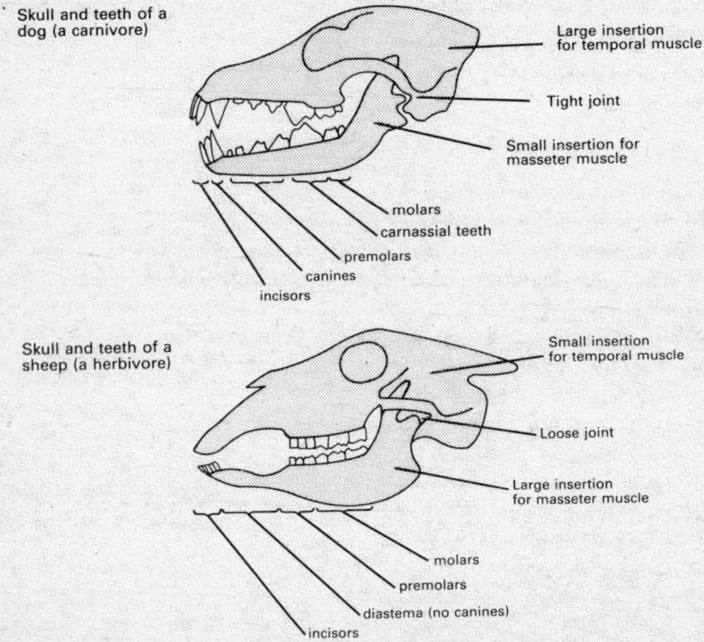

Fig. 5.17. *Comparison of herbivore and carnivore jaws and teeth*

(a) Pointed incisors are used for nibbling meat from bones.

(b) Large, pointed canines are used to kill the prey, by piercing veins in the neck.

(c) Premolars and molars are flattened along the line of the jaw and have a scissor-like action for slicing flesh and crunching bones. The **carnassial teeth** are specially adapted for this. Sliced meat is swallowed without chewing.

(d) The tight joint between skull and jaw and the position of the masseter muscle restricts side to side movement of the jaw.
(e) The temporal muscle (which moves the jaw up and down) is large, producing a powerful bite.
3. A **sheep** is a **herbivore**. Its dentition is specially adapted to crop and grind tough vegetation. Refer to Fig. 5.17 as you go through the following points:
(a) The flattened incisors of the bottom jaw meet against a horny pad on the top jaw and are used for tearing or pulling out grass.
(b) The canines are absent. A large gap, the **diastema**, serves to separate newly cropped grass from that being ground up by the cheek teeth.
(c) The cheek teeth are large and ridged, the ridges on the bottom jaw fitting into those on the top. The two sets of teeth grind against each other in a circular motion.
(d) The cheek teeth grow continuously because of the considerable wear they receive.
(e) The masseter muscle (which moves the jaw from side to side) is large. The loose joint between jaw and skull allows a sideways rocking action as grass is chewed.
(f) The temporal muscle (which moves the jaw up and down) is small.

Q. (a) Make a labelled diagram to show the structure of a mammalian tooth.
(b) In what ways do teeth of mammals differ from those of other vertebrates?

Dental disorders

Two types of dental disorder can lead to loss of teeth: **dental caries (tooth decay)** and **periodontal (gum) disease**. Both disorders involve an accumulation of a surface film on the teeth called **plaque**.

Plaque is a mixture of saliva, food and bacteria (see page 294). The bacteria feed on the food in plaque and release acids which attack the teeth and gums.

Dental caries (tooth decay) is caused by the action of bacteria which break down sugar in the plaque to form acid. The acid dissolves the mineral salts first in the tooth enamel and then in the dentine, forming a hole (**cavity**). If left untreated, the pulp may become infected and pain (toothache) results. Eventually an abscess may form in the bone underlying the tooth. At this stage, the tooth must be removed and the abscess treated.

Repair of a tooth damaged by caries consists of removing the decayed area using a drill and replacing it with a filling usually of silver and tin mixed with copper and zinc.

Periodontal (gum) **disease** is caused by accumulation of plaque around the neck of the tooth. Poisons released by bacteria in the plaque irritate the gums and cause swelling. The gums may bleed slightly when the teeth are brushed. The gum retreats leaving more of the tooth exposed, (getting 'long in the tooth' is a sign of early periodontal disease rather than simply increasing age). Eventually poisons released by the bacteria destroy the fibrous membrane holding the tooth in the socket and the tooth falls out. In Britain, many more teeth are lost in this way, through gum disease, than through tooth decay.

Care of teeth

Tooth decay and gum disease can be effectively prevented by a combination of the following measures:

1. restricting the amount of sugar in the diet: The presence of sugar in the mouth provides food for the acid-producing bacteria. Chewing gum and sweets between meals is best avoided. Other sugar-rich foods such as cake and biscuits are best taken only at meal times.
2. removal of plaque by regular cleaning: Ideally, the teeth should be cleaned thoroughly with a toothbrush and toothpaste after each meal. Failing this, you should brush your teeth last thing at night and first thing in the morning or just after breakfast. The brushing action should be up and down, rather than side to side, to get at food between the teeth. Brush from the gum downwards for the top teeth and upwards for the bottom teeth. This helps prevent food being pushed back to the neck of the tooth. Effective brushing takes at least a full two minutes. The close gap between teeth can be cleaned using dental floss.
3. the use of fluoride: Fluoride is a natural component of tooth enamel. It strengthens the enamel and makes it more acid-resistant. The addition of one part per million (1 ppm) of fluoride to drinking water has been shown to reduce the incidence of tooth decay and hence fillings in children. Some water authorities add fluoride to the water, others do not. The use of fluoride toothpaste can be beneficial.
4. regular visits to the dentist: Visit the dentist every six months for a check-up from an early age. The dentist can help improve badly shaped teeth which would otherwise trap plaque in places which are difficult to brush. The dentist will also fill cavities and remove deposits of hardened plaque known as **calculus** (tartar).

5.15. *The Alimentary Canal of Carnivores and Herbivores*

Just as the dentition and jaw structure of mammals differ according to diet, so the basic plan of the gut is modified.

1. **Carnivores** have a relatively short gut because their food is protein-rich and easy to digest.
2. **Herbivores** have a diet rich in plant fibre (indigestible to omnivores and carnivores) and need special adaptations to utilize this source of food. Mammals cannot make their own cellulose-digesting enzymes, but herbivorous mammals have developed an association (**symbiosis**) with certain micro-organisms which provide the necessary enzymes. A herbivore's intestines are long as plant material is difficult to digest.

(a) **Ruminants**, e.g. cows and deer, have a large outgrowth of the stomach, the **rumen**. Here, symbiotic bacteria produce cellulose-digesting enzymes. The freshly ingested food is passed to the rumen where it is 'fermented' (partially digested). It is later regurgitated and thoroughly chewed ('chewing the cud') before passing into the true stomach.

(b) **Non-ruminants**, e.g. horses and rabbits, do not have a rumen, but culture their cellulose-digesting bacteria in the caecum and appendix. Rabbits 'refect', eat their faeces, thus passing food through the gut twice.

All these adaptations of herbivores serve to increase their ability to digest fibre-rich food.

5.16. *Feeding Methods in Holozoic Organisms*

Most animals show **extracellular digestion**: food is digested outside cells, in a gut or outside the organism.

Single-celled animals, e.g. *Amoeba* (page 315) show **intracellular digestion**: food is digested within the cell.

In general, animals obtain their food in one of three forms:

1. as large solids
2. as solids suspended in liquid
3. as liquids.

1. **Large solids** must be broken down by biting and chewing into particles which are small enough to be ingested and swallowed.

Those insects (page 319) which have chewing mouthparts include the locust (a herbivore), the cockroach (an omnivore), and the ground beetle (a carnivore).

Those mammals (page 329) which have chewing mouthparts include the sheep (a herbivore), humans (omnivores) and the dog (a carnivore).
2. **Suspended solids** need to be filtered out of the liquid before they can be ingested. Usually a sieve-like apparatus is used to do this. As suspended organisms are usually very small, vast quantitities of water must be filtered to provide an adequate food supply. This method of feeding (filter-feeding) is found in certain aquatic organisms, such as the herring (page 323).
3. Those animals which feed on **liquids** have sucking mouthparts. In the housefly, the mouthparts form a hollow tube (the **proboscis**) which is used for releasing saliva as well as sucking up food. The saliva contains digestive enzymes which convert solid organic matter to a semi-digested liquid. This liquid can then be drawn up the proboscis and into the stomach by muscular contraction.

The mouthparts of the mosquito are for piercing as well as sucking. The proboscis is pointed and in the female is used to pierce mammalian skin to obtain blood, and in the male, to pierce plants to obtain phloem.

5.17. *Saprophytic Nutrition*

Saprophytes play a major role in the cycling of chemicals in nature; see decay bacteria and fungi (Figs. 14.1 and 14.2, pages 269–70) and nitrogen-fixing and nitrifying bacteria (page 270 and Fig. 14.2). *Mucor* (page 301) is an example of a saprophytic fungus.

5.18. *Parasitic Nutrition*

For examples of parasites, see viruses (page 292), disease-causing bacteria (page 294) and the fungus *Phytophthora* (page 302).

Definitions

Ingestion
Digestion
Absorption
Assimilation
Egestion
Secretion

Key Words

Autotrophic
Heterotrophic
Holozoic
Saprophytic
Parasitic
Host
Chloroplast
Midrib
Vein
Lamina
Palisade mesophyll
Spongy mesophyll
Air space
Epidermis
Cuticle
Guard cell
Stomata
Xylem
Phloem
Variegated
Light reaction
Dark reaction
Limiting factor
Wilting
Water culture

Carnivore
Herbivore
Omnivore
Balanced diet
Peristalsis
Mucus
Buccal cavity
Oesophagus
Stomach
Duodenum
Liver
Pancreas
Ileum
Caecum
Appendix
Colon
Rectum
Anus
Salivary amylase
Bolus
Hydrochloric acid
Pepsin
Rennin
Chyme
Pyloric sphincter

Emulsify
Lipase
Trypsin
Erepsin
Sucrase
Maltase
Lactase
Villi
Hepatic artery
Hepatic vein
Bile duct
Gall bladder
Pulp cavity
Homodont
Heterodont
Incisor
Canine
Premolar
Molar
Diastema
Carnassial teeth
Extracellular digestion
Intracellular digestion

Questions

1. (a) Give a balanced equation which summarizes the process of photosynthesis.
(b) Describe in detail an experiment which would demonstrate that light is needed for photosynthesis.
(c) What is the importance of photosynthesis to animals?

2. State four *environmental conditions that enable photosynthesis to occur*

3. For each of the following food materials describe one *test you would use to show its presence:*

(i) a reducing sugar
(ii) starch
(iii) a protein
(iv) a fat

4. (a) List the food requirements of mammals.
(b) In what ways are herbivorous and carnivorous mammals adapted to their respective diets?

5. The most important function of the villi lining the small intestine of a mammal is to
A. increase the surface area
B. push the food along
C. secrete increased quantities of enzymes
D. trap bacteria
E. direct the flow of blood.

6. Vitamin A is essential in man's diet to prevent
A. pellagra
B. scurvy
C. beri-beri
D. rickets
E. night blindness.

6. *Respiration, Gas Exchange and Breathing*

Definition: Respiration *is the breakdown of organic substances, within cells, to release energy.*

Respiration occurs in all living cells since every cell needs a supply of energy to remain alive.

The term respiration should not be confused with gas exchange or breathing (ventilation).

Definition: Gas exchange *is the exchange of oxygen for carbon dioxide between an organism and its surroundings.*

Gas exchange usually accompanies respiration. The surface at which it occurs is called a respiratory or gas exchange surface.

Breathing or ventilation refers to the movements which cause a renewal of air or water at a respiratory surface. Breathing occurs in larger animals such as the larger arthropods and in vertebrates.

Definition: Breathing (ventilation) *is the maintenance of a flow of air or water over a respiratory surface.*

The term respiration is sometimes referred to as **internal**, **cell** or **tissue respiration**, and gas exchange and breathing grouped under the term **external respiration**. This use of terms can lead to confusion. The more modern naming system used in this book is simpler and clearer.

You must be absolutely clear about these terms. They often form the basis of long-answer questions. Clearly if you have got their meaning wrong you are in trouble.

The key feature of respiration is that it releases energy. Organisms need this energy to carry out the following essential processes:

1. Anabolic metabolism in which energy is used to make complex substances from simpler ones.
2. Muscle contraction to do mechanical work.
3. Generation of nerve impulses.
4. Active transport of substances into and out of cells.
5. In birds and mammals, heat energy is used to maintain a high body temperature.

This released energy is stored in the cells as a chemical, adenosine triphosphate (ATP). When the cell requires energy ATP is converted to adenosine diphosphate (ADP) and energy is released. There are two forms of respiration, **aerobic** and **anaerobic**.

6.1. *Aerobic and Anaerobic Respiration*

Aerobic respiration is the commonest form of respiration and can only occur in the presence of oxygen. A simplified equation for aerobic respiration is given below:

$$\underset{\text{glucose}}{C_6H_{12}O_6} + \underset{\text{oxygen}}{6O_2} \rightarrow \underset{\text{carbon dioxide}}{6CO_2} + \underset{\text{water}}{6H_2O} + \text{energy (2,880 kJ)}$$

As you can see, the glucose molecule is broken down into much smaller, simpler molecules. It is this breaking of chemical bonds that releases energy.

Anaerobic respiration does not need oxygen to occur but it does not release as much energy as aerobic respiration.

In bacteria and certain fungi, the end product of anaerobic respiration is alcohol. A simplified equation for this process is given below:

$$\underset{\text{glucose}}{C_6H_{12}O_6} \rightarrow \underset{\text{alcohol}}{2C_2H_5OH} + \underset{\text{carbon dioxide}}{2CO_2} + \text{energy (210 kJ)}$$

Turning glucose into alcohol is called **fermentation** (see section 6.2). This process only releases 210 kilojoules of energy compared to 2,880 kilojoules from aerobic respiration. This is because fewer chemical bonds are broken during anaerobic respiration.

During vigorous exercise the muscles of animals sometimes have to use anaerobic respiration, when they are not receiving oxygen as quickly as they are using it. The end product is lactic acid:

$$\underset{\text{glucose}}{C_6H_{12}O_6} \rightarrow \underset{\text{lactic acid}}{2CH_3CH(OH)COOH} + \text{energy (150 kJ)}$$

This lactic acid builds up in the muscles and eventually causes fatigue. After exercise, the lactic acid must be removed from the muscles. This is achieved by transporting the lactic acid to the liver and using oxygen to convert it back to glucose. This means that for a while after exercise the body still requires more oxygen than usual, to deal with the lactic acid. This is known as an **oxygen debt** and is the reason why it is necessary to go on breathing heavily for a short time after strenuous exercise.

6.2. *Using Anaerobic Respiration*

The ability of certain fungi to turn sugars into alcohol has been exploited by humans for many years in the manufacture of alcoholic drinks. The best known of these fungi is **yeast**.

To make wine, yeast is used to ferment the sugars in grapes. Originally the process depended on the presence of wild yeast on the skins of the grapes. Nowadays, most wines are made using special strains of yeast that are added to the grapes.

To make beer, hops and barley are acted on by yeast; using apples produces cider.

In Brazil, alcohol produced by fermentation of sugar cane is used as fuel for cars.

Yeast is also used in the manufacture of bread. When yeast is added to the dough carbon dioxide is given off and causes the dough to rise.

6.3. *Respiration Experiments*

A demonstration of respiration in material is one indication that the material is living.

Measurements of the rate of respiration give some idea of the rate of metabolic activity.

If we look at the equation for respiration on page 92 we see that an organism respiring aerobically should:

1. use up carbohydrate
2. use up oxygen
3. give off carbon dioxide
4. produce water or water vapour
5. release energy.

If we can show that one of the above is occurring we can infer that the organism is living and respiring. In practice, 1, 2, 3 and 5 are suitable tests for respiration. 4 is not an effective test because:

(a) Non-living as well as living material may give off water by evaporation.

(b) Only a fraction of the water given off by an organism comes from respiration; most has originally been absorbed.

Of the four possibilities remaining, 3 and 5 are the easiest to demonstrate, and questions on these respiration experiments are commonly included in exams.

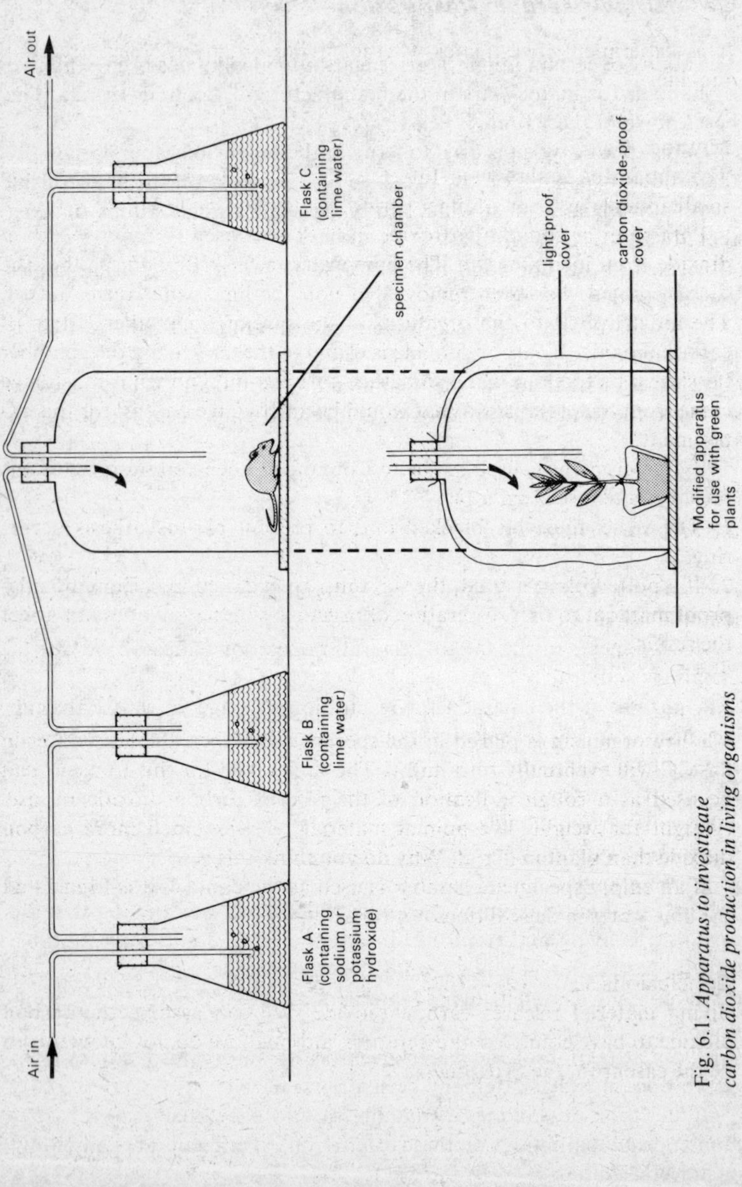

Fig. 6.1. *Apparatus to investigate carbon dioxide production in living organisms*

Investigation of carbon dioxide production by living organisms

This experiment investigates whether living organisms release carbon dioxide.

Method

The apparatus is shown in Fig. 6.1. The experiment is effective using small animals or, over a longer period, using plant material.

Potassium or sodium hydroxide in flask A is used to remove carbon dioxide from incoming air. The lime water in flask B confirms that the carbon dioxide has been removed (if not, the lime water turns milky). The air delivered to the organism in the specimen chamber is free of carbon dioxide. If carbon dioxide is found in the air leaving the chamber (as shown by the lime water in flask C turning milky), then it must have come from the organism. What would be an effective control for this experiment?

If a green plant is used as the test organism, then certain precautions must be taken (see Fig. 6.1):

1. The vessel must be 'blacked out' to prevent photosynthesis occurring.
2. If a potted plant is used, the soil must be enclosed in carbon dioxide-proof material so that respiration of organisms in the soil does not affect the result.

Results

If a live organism is placed in the specimen chamber, the lime water in flask C will eventually turn milky. The time taken for this to occur can be used as a rough indication of the rate of carbon dioxide release. (Weight for weight, live animal material releases much more carbon dioxide than plant material. Why do you think this is so?)

If an empty specimen chamber is used as the control, it is found that the lime water in flask C does not turn milky.

Conclusions

Living material releases carbon dioxide. We can assume the carbon dioxide to have come from respiration, although we do not know this to be the case from the experiment.

Investigation of carbon dioxide production by anaerobic organisms

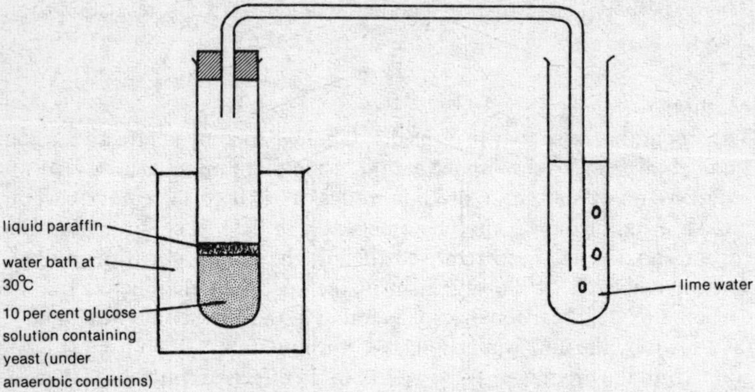

Fig. 6.2. *Apparatus to investigate carbon dioxide production under anaerobic conditions*

A modification of the above experiment can be used to show that anaerobic organisms such as yeast can produce carbon dioxide in the absence of oxygen. The experiment is set up as shown in Fig. 6.2.

The water is boiled to remove carbon dioxide before the yeast and glucose are added. The layer of liquid paraffin prevents oxygen entering the liquid. A control apparatus, identical but without the yeast, is also set up.

After several hours the lime water in the test apparatus turns milky, while that in the control apparatus remains clear. The results, taken together, indicate that yeast liberates carbon dioxide under anaerobic conditions. In addition, the presence of alcohol (a product of anaerobic respiration) can be detected by smell.

Using sodium hydrogencarbonate indicator

The lime water in this experiment can be replaced with sodium hydrogencarbonate indicator to indicate the presence of carbon dioxide. This indicator is purple in the presence of low levels of carbon dioxide, such as would be found normally in the air. As carbon dioxide levels increase the indicator changes to yellow.

Investigation of energy release in germinating seeds

This experiment investigates whether live seeds release energy. Heat production is a good indication of energy release.

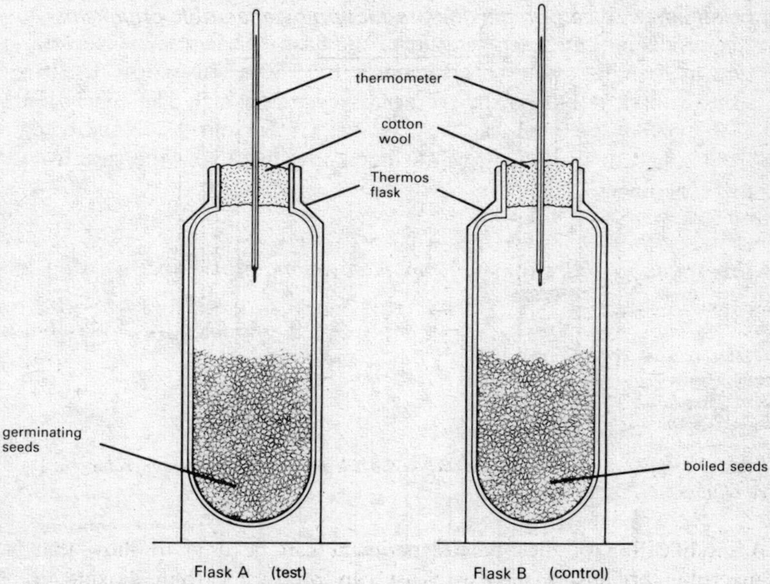

Fig. 6.3. *Apparatus to investigate heat production in germinating seeds*

Method

A suitable experiment is shown in Fig. 6.3. Wheat seeds are used as the experimental material. Two batches of seeds of equal weight are used. One batch is killed by placing in boiling water for 10 minutes. These seeds will be used for the control.

Both batches are rinsed with hypochlorite solution to prevent bacterial or fungal growth. The live seeds are then placed in flask A and the dead seeds in flask B. Thermometers are inserted and the mouths of the flasks plugged. The temperature within the flasks is recorded at one-day intervals.

Results

Typical results are as follows:

	Temperature °C	
Day	*Flask A*	*Flask B*
1	21	20
2	23	20
3	25	19
4	26	20

Conclusion

The results indicate a temperature rise in flask A, whereas the temperature in flask B remains nearly constant. Heat is released by the live seeds in flask A but not by the dead seeds in flask B. The conclusion is that living seeds release heat. This can be assumed to have come from respiration although we do not know this to be the case from this experiment.

Investigation of gas (oxygen) uptake by living organisms

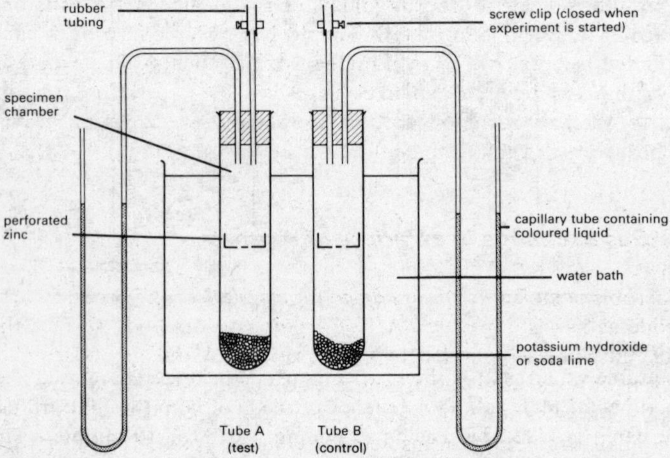

Fig. 6.4. *Apparatus to investigate oxygen uptake in living organisms*

The experiment investigates whether living organisms take up oxygen.

Method

Oxygen uptake cannot be measured directly. However, if an organism is respiring aerobically the oxygen it takes up will be converted to carbon dioxide. If the carbon dioxide is absorbed by soda lime or potassium hydroxide in a closed container, any drop in gas volume can be attributed to oxygen uptake by the organism.

A suitable apparatus is shown in Fig. 6.4. Germinating seeds or small animals are suitable test material. In the case of seeds, dead seeds can be used for the control. The apparatus is immersed in a water bath to minimize temperature changes which might affect the gas volume readings.

The apparatus is left in the water bath for five minutes with the screw tops open, and then the screw tops are closed and the experiment begun.

Results

After about half an hour the level of coloured liquid in the capillary tubing of A will typically rise on the side nearest the specimen chamber. In the capillary tubing attached to B little or no change in the level of the liquid is observed.

Conclusions

The results indicate a drop in volume of gas in tube A but little or no change in the volume in tube B. The drop in volume in tube A can be attributed to uptake of oxygen by the live specimen(s). The experiment shows that a living organism takes up oxygen.

NOTE. We have assumed that the gas taken up is oxygen but we have not positively shown this to be the case.

6.4. *Gas Exchange in a Variety of Animals*

All aerobic organisms require oxygen for respiration and produce carbon dioxide as a waste product. At some point on the body surface these gases must be exchanged between the organism and the environment. Oxygen diffuses into the organism and carbon dioxide diffuses out, along a concentration (diffusion) gradient (see page 38).

In small simple organisms such as *Amoeba* (page 315) all parts of the body are only a short distance from the external environment and gas exchange occurs over the entire body surface. As organisms increase in size and complexity the inner parts of the body are too far away from the outside to get their oxygen and get rid of carbon dioxide by diffusion alone. Also the surface area to volume ratio decreases and diffusion cannot proceed at a high enough rate to supply the tissues (see Fig. 6.5). A specialized respiratory surface, e.g. the gill, is required, and a transport system to deliver gases to and from all cells in the body.

All respiratory surfaces have the following features which permit diffusion:

1. They are thin (to allow gases to diffuse rapidly through).
2. They are moist (gases must be in solution before they can pass across a cell membrane).
3. They have a large surface area (to maximize diffusion).

As organisms increase in size, the body surface increases by the square of its dimensions, whereas the volume increases by the cube. The result is that the larger the organism, the smaller the surface area : volume ratio. This can be illustrated by looking at cubes of increasing size:

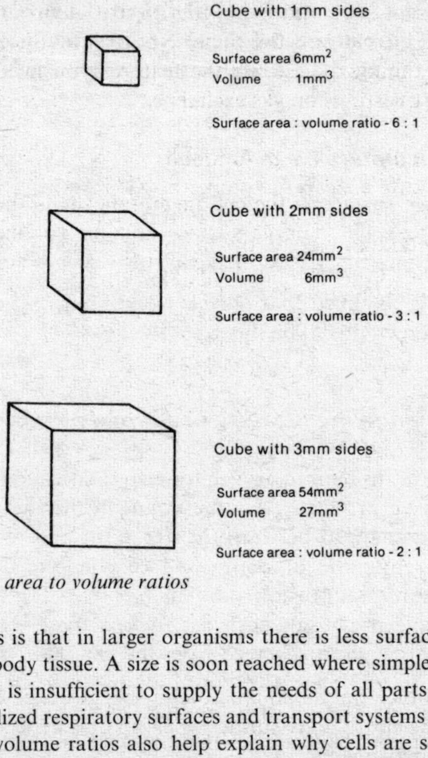

Fig. 6.5. *Surface area to volume ratios*

The effect of this is that in larger organisms there is less surface area to supply each volume of body tissue. A size is soon reached where simple diffusion across the body surface is insufficient to supply the needs of all parts of the body. At this stage, specialized respiratory surfaces and transport systems are required.

Surface area : volume ratios also help explain why cells are so small. As they increase in size, the surface area which supplies food and oxygen to, and removes wastes from, each volume of protoplasm, decreases. A limiting size is soon reached. Beyond this all parts of the cell are unable to obtain supplies and get rid of wastes quickly enough to support life.

4. They are permeable to gases.

In addition, in larger organisms and those which have a high energy demand, more oxygen is required and the respiratory surface must satisfy the following conditions:

5. It must be well ventilated.
6. It must be richly supplied with blood, which usually contains a

respiratory pigment, e.g. haemoglobin. This increases the amount of oxygen that can be carried in the blood.

NOTE. Insects are an exception (see page 102).

You need to be familiar with the gas exchange surfaces of a range of animals of different sizes and living in different environments. In each case the basic principles are the same. Notice how the above features apply to each example. These are the features you need to point out when answering questions on gas exchange.

Gas exchange in protozoa, e.g. Amoeba

The gas exchange surface is the cell membrane. It is thin, moist (surrounded by water), has a large surface area to volume ratio and is permeable to respiratory gases. Gas exchange occurs across the entire cell membrane (Fig. 6.6).

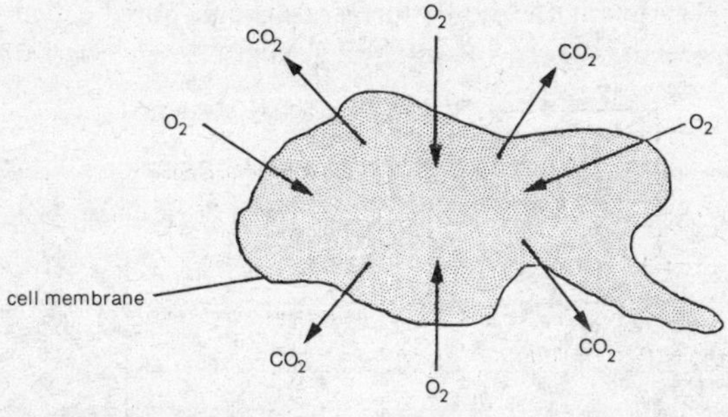

Fig. 6.6. *Gas exchange in* Amoeba

Similarly, in other simple aquatic organisms gas exchange occurs by diffusion over the entire body surface. Please note, these animals do not show breathing movements.

Gas exchange in the earthworm

In the earthworm, as in *Amoeba*, the entire body surface is the gas exchange surface. The earthworm is much larger, however, and a transport system is needed to ensure that internal tissues receive enough oxygen. The skin epidermis is one cell thick and is kept moist by mucus secreted from gland cells (see Fig. 6.7). Oxygen dissolves in the surface moisture of the epidermis and diffuses into blood capillaries just below. Carbon dioxide diffuses out in the opposite direction. The blood system, containing dissolved haemoglobin, conveys oxygen to all parts of the body.

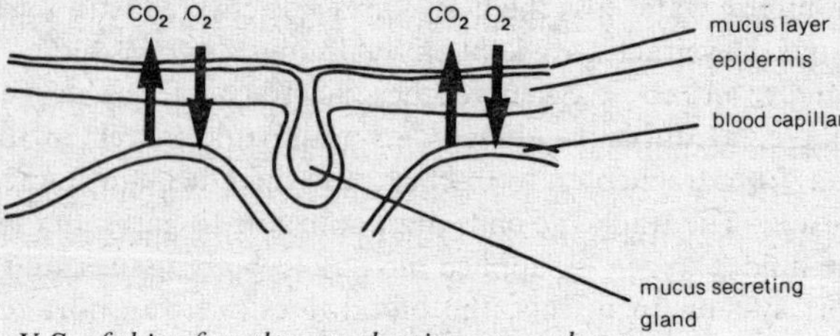

Fig. 6.7. *V.S. of skin of earthworm showing gas exchange* (NOTE. *Frogs have a similar arrangement*)

Gas exchange in insects

Gas exchange in insects occurs at the end of small air tubes called **tracheoles** (see Fig. 6.8). These are connected to larger air passages, **tracheae** (singular trachea), which are supported by rings of chitin. The tracheae open out at the body surface through pores called **spiracles**. The whole system is called the **tracheal system**.

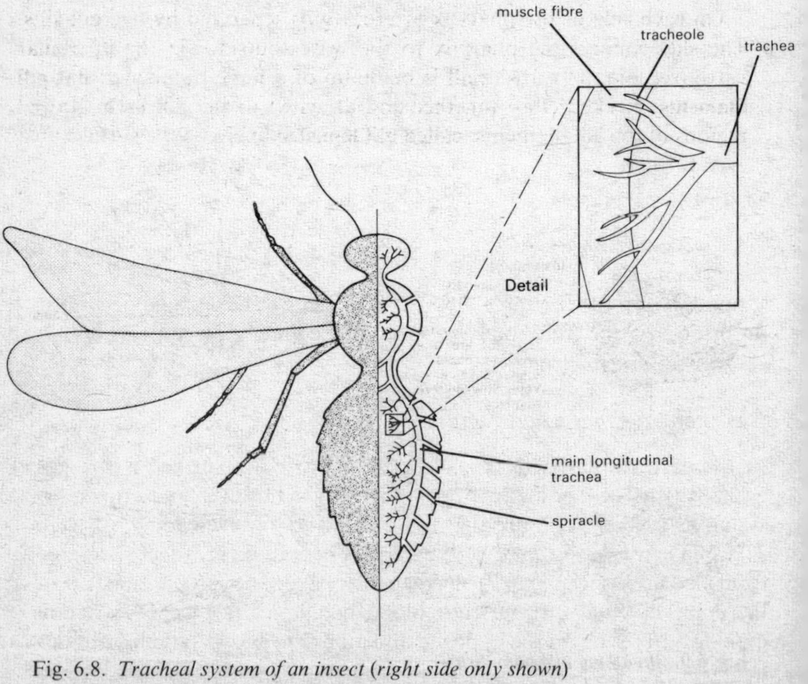

Fig. 6.8. *Tracheal system of an insect* (*right side only shown*)

To minimize water loss the body surface is covered with a waterproof cuticle and the spiracles are kept closed as much as possible.

In smaller insects, oxygen reaches the tracheole endings, from the atmosphere, by diffusion alone. Carbon dioxide travels in the reverse direction, from tracheoles, to tracheae and then through spiracles, again by diffusion. The tracheole ends are permeable to gases and are bathed in tissue fluid. Oxygen is supplied *directly* to body tissues, and *not* via a transport system. In insects, the blood does *not* transport respiratory gases.

Some insects show pumping movements of the abdomen which squeeze air in and out of the tracheal system. This ventilation (breathing) aids diffusion and increases the rate of gas exchange.

Gas exchange and breathing in fish

In fish, the site of gas exchange is the epithelium of gill filaments. It is moist, thin-layered, has a large surface area and is richly supplied with blood capillaries. It is also well ventilated.

On each side of the pharynx are four gills separated by five gill slits. The slits connect the pharynx to the water outside via the opercular valve (see Fig. 6.9). Each gill is made up of a large number of flat **gill filaments**, stacked close together and attached to the **gill arch**. Raised regions of the gill filaments, called **gill lamellae**, increase the surface area even further.

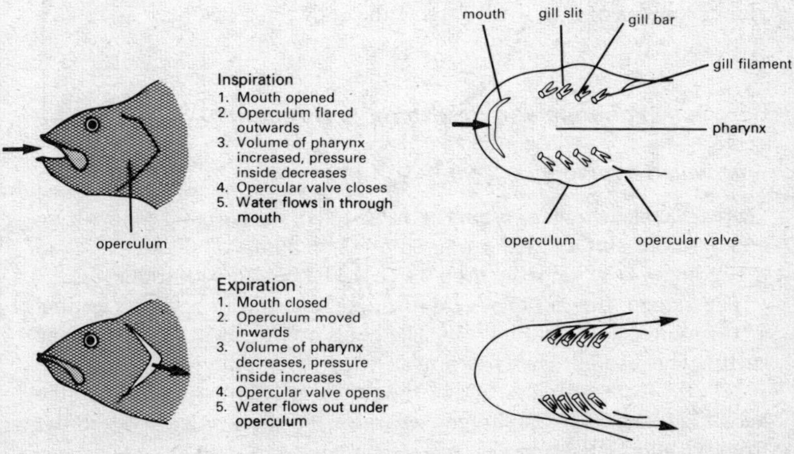

Fig. 6.9. *Breathing in a bony fish*

A flow of oxygen-rich water over the gills is maintained by changes in volume of the pharynx which in turn generate pressure changes inside. During inspiration, the mouth is open and muscular action increases the volume of the pharynx, decreases the pressure inside it, and water flows in through the mouth. During expiration, the mouth is closed and muscular action decreases the volume of the pharynx, increases the pressure inside it, and water flows out through the open opercular valve.

The gas exchange surface, the gill epithelium, is only one cell thick. Oxygen dissolved in the water diffuses into the blood and combines with haemoglobin. The oxygen is carried in the blood to all parts of the body. Carbon dioxide in the blood diffuses out through the gas exchange surface and into the water. The water expelled under the operculum is rich in carbon dioxide.

Gas exchange in the frog

The adult frog has three respiratory surfaces: the lining of the mouth, the lungs and the skin. Gas exchange across the mouth lining is used when the frog is on land and resting. The lining is ventilated by visible throat movements. The lungs are small and simple and are used when the frog is active on land. The skin is the main respiratory surface and is used at all times, including swimming or hibernation. It has a thin epidermis covered with mucus, and supplied by a network of capillaries, as in the earthworm (page 102). A blood transport system containing haemoglobin conveys oxygen to all parts of the body.

6.5. *Gas Exchange and Breathing in Humans*

The human respiratory system

Gas exchange in humans occurs at the **alveoli** of the **lungs**. The lungs have an enormous surface area produced by the millions of alveoli (air sacs) they contain. The lungs are deep inside the body to reduce water loss.

The human respiratory system (the lungs and the air passages leading to them) are shown in Fig. 6.10. The lungs occupy most of the **thoracic cavity**. This airtight cavity is protected by the sternum and ribs (page 192) at the front and sides, and by the vertebral column (page 193) at the back. The muscular diaphragm separates the thoracic cavity (thorax) from the abdominal cavity.

Notice the lungs are separated from the ribs by a space, the **pleural**

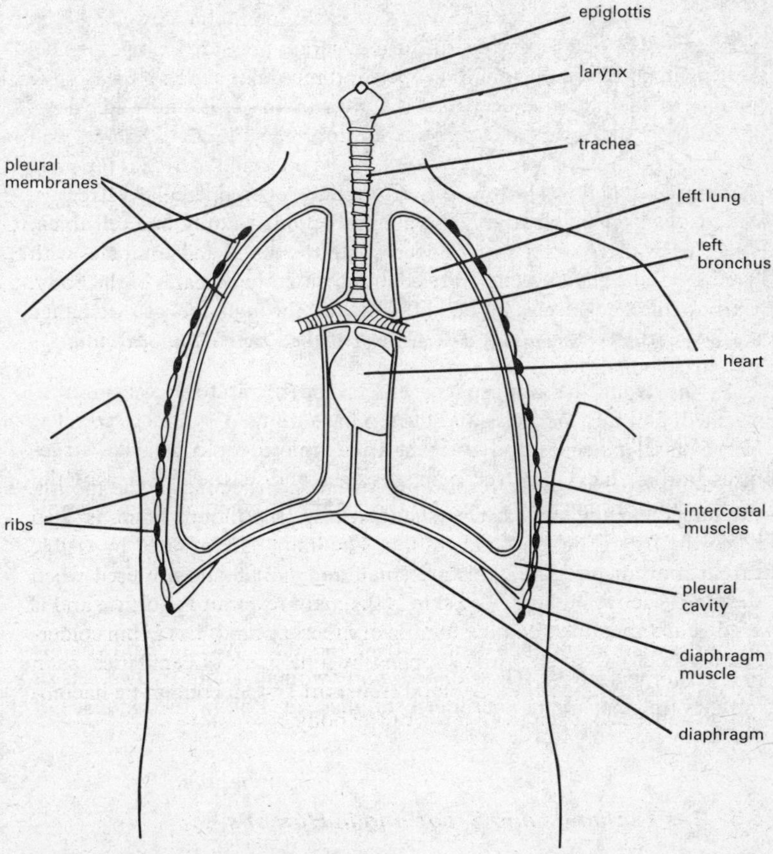

Fig. 6.10. *The human respiratory system*

cavity, bordered on inside and outside by the **pleural membranes**. These membranes prevent the lungs being damaged by the ribs. The pleural cavity contains a fluid which lubricates the membranes so they move past each other easily.

On breathing in (inspiration) air reaches the alveoli by first entering the mouth or nostrils, passing through the **pharynx**, down the **trachea** (windpipe), and then through the two **bronchi** and into thousands of **bronchioles** in the lungs. The bronchioles end in groups of alveoli looking rather like microscopic bunches of grapes (see Fig. 6.11). Each alveolus is covered by capillaries and it is here that gas exchange takes place.

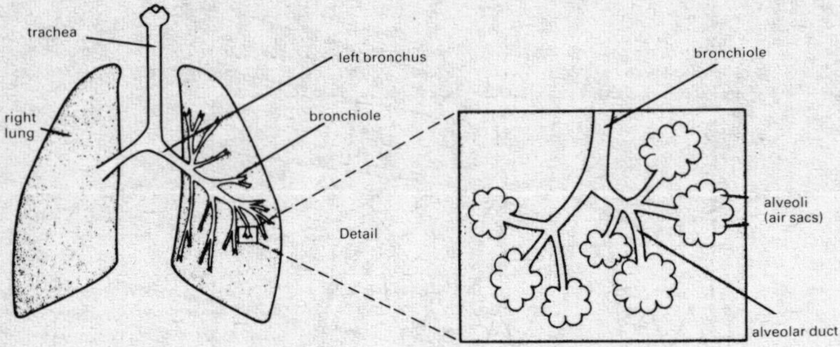

Fig. 6.11. *Detail of human lung showing smaller air passages and alveoli*

As air from the atmosphere enters the respiratory system it is warmed, and dust particles and bacteria are trapped by mucus secreted in the nasal passages and trachea. Cilia (microscopic hair-like structures but *not* hairs) line these air passages and as they beat, waft the mucus and trapped particles to the pharynx where they are swallowed.

Breathing

Air is drawn into and then expelled from the lungs by changes in volume of the thoracic cavity. These changes are brought about by two sets of muscles; the **diaphragm** and **intercostal muscles**. Follow the sequence of events by referring to Fig. 6.12.

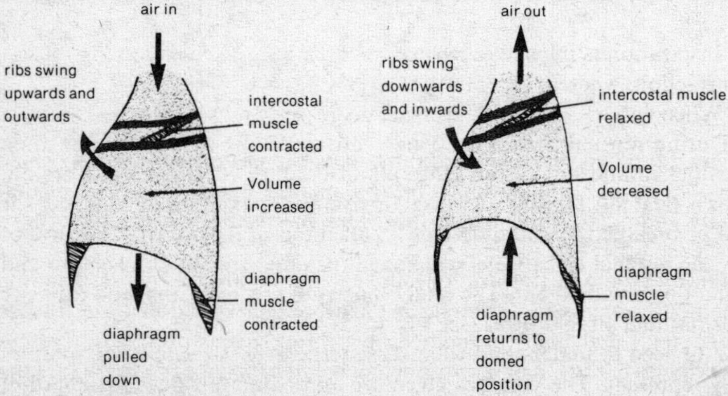

Fig. 6.12. *Breathing in humans*
(NOTE. *Only 2 ribs are shown*)

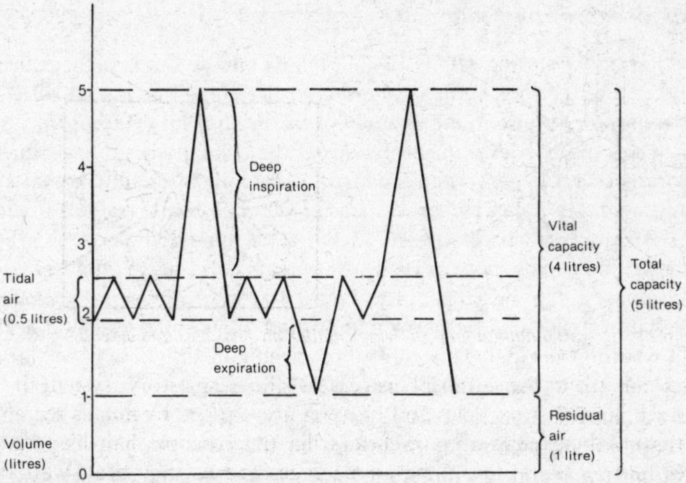

Fig. 6.13. *Volumes of air in the lungs*

Inspiration (breathing in): diaphragm and intercostal muscles contract, thorax increases in volume, pressure inside thorax decreases, and air moves into the lungs and they inflate.

Expiration (breathing out): diaphragm and intercostal muscles relax, thorax decreases in volume, pressure inside thorax increases, and air moves out of the lungs and they deflate.

It is essential to get these sequences right. There are two points to notice:

1. Inspiration is the active part of the breathing cycle, involving muscle contraction, whereas expiration is largely passive.
2. When volume increases, pressure decreases, and vice versa.

During normal breathing (what you are doing now) an adult man breathes in and out of the lungs about 0.5 litres of air in each breath. This is **tidal air**. During very deep breathing a maximum of 4 litres of air can be breathed in and out. This is called the **vital capacity**. One litre of air, the **residual air**, always remains in the lungs trapped by friction and cannot be breathed out. The **total capacity** of the lungs, the vital capacity plus residual air, is 5 litres (see Fig. 6.13).

You need to understand what these terms refer to, and what volumes they represent. The volumes given are approximate values for an adult man.

Control of breathing rate

The normal breathing rate is 15–20 breaths/minute but is higher during and just after periods of physical exercise. Breathing rate is controlled by the **respiratory centre** in the medulla of the brain. This centre is sensitive to carbon dioxide level in the blood. If the level increases, e.g. during vigorous exercise, nerve impulses from the brain to the intercostal and diaphragm muscles cause an increase in ventilation rate (rate of breathing). This in turn speeds up gas exchange to remove the excess carbon dioxide (and supply more oxygen). At the same time nerve impulses from the brain speed up heartbeat rate and the transport of respiratory gases.

As carbon dioxide levels return to normal so breathing rate and heartbeat return to normal. This control of carbon dioxide level in the blood is an example of **homeostasis** (see pages 147 and 179).

An experiment to compare breathing rates

Breathing rates can be compared before and after exercise very simply. The breathing rate is counted by watching the rise and fall of the chest. The breathing rate should be faster after exercise. You should be aware that this experiment is not completely accurate because the breathing rate alters when the subject is thinking about breathing.

Gas exchange at alveoli

There are over 500 million alveoli in the human lungs. Alveoli have all the features we have come to expect of a respiratory surface in a large, active animal. They are thin, moist, have a large surface area, and are freely permeable to gases. In addition, they are well ventilated and have a rich blood supply; both are features which maintain a steep concentration gradient.

The air in the alveolus (alveolar air) has a higher concentration of oxygen than the arriving blood in the overlying capillary. Oxygen dissolves in the moisture lining the alveolus and diffuses across two layers of cells (the lining ot the alveolus and the lining of the blood capillary) and into the blood. Here it diffuses into red blood cells (page 117) and combines with **haemoglobin** to form oxyhaemoglobin.

Haemoglobin is a respiratory pigment and because of its attraction for oxygen, enables the blood to carry large amounts of the gas. Haemoglobin gives blood its red colour.

Carbon dioxide is carried in the blood as the **bicarbonate ion**. Blood arriving at the alveolus is rich in carbon dioxide. The concentration of

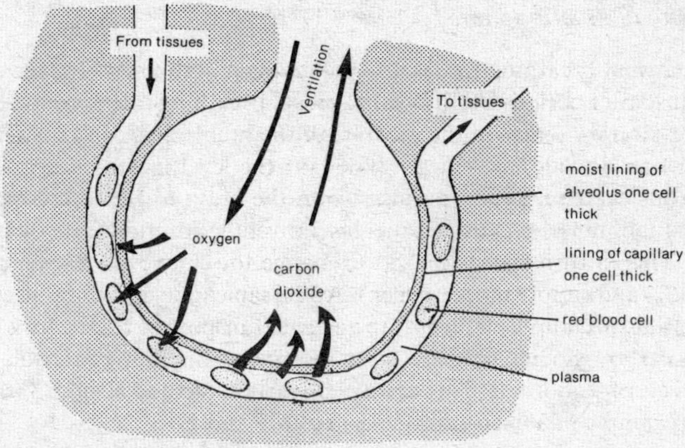

Fig. 6.14. *Gas exchange at an alveolus*

Table 6.2. *Composition of gases in breathed air*

	Inspired air	*Expired air*	*Alveolar air*
Oxygen	20%	16%	14%
Carbon dioxide	0.04%	4%	6%
Nitrogen	79%	79%	80%
Water vapour	Variable	Saturated	Saturated

carbon dioxide in the alveolar air is lower. At the alveolus the bicarbonate breaks down to release carbon dioxide, which diffuses from the blood into the alveolus.

The blood, now rich in oxygen (oxygenated) and low in carbon dioxide is carried away to supply tissues throughout the body. The path taken by blood is shown in Fig. 7.2.

The composition of alveolar air remains nearly constant (see Table 6.2) despite the continual loss of oxygen to the blood and gain of carbon dioxide. Replacement oxygen diffuses across from inspired air and carbon dioxide is lost to expired air (see Fig. 6.14).

An experiment to compare exhaled and inhaled air

This experiment makes use of the apparatus shown in Fig. 6.15.

If you breathe gently through the tube indicated, air flows through the apparatus in the direction indicated by the arrows. The lime water in flask X therefore receives air from outside, equivalent to the air we

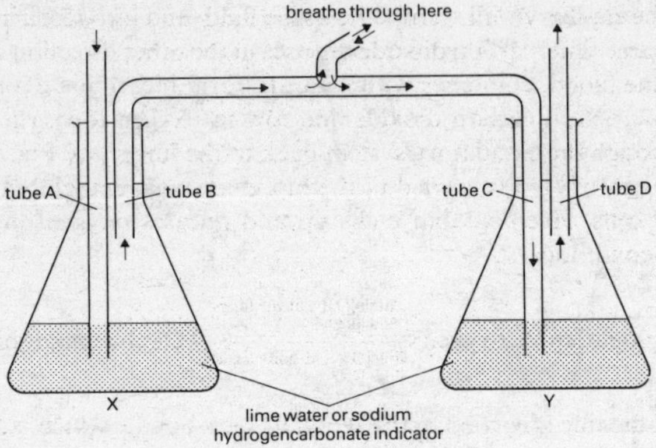

Fig. 6.15. *Comparing exhaled and inhaled air*

breathe in. The lime water in flask Y receives exhaled air. The lime water in flask Y goes cloudy before the lime water in flask X, showing that exhaled air contains more carbon dioxide than inhaled air.

NOTE. Sodium hydrogencarbonate indicator can be used instead of lime water.

Supplying tissues with oxygen and removing carbon dioxide

On arrival at the site of respiring tissues in the body, the oxyhaemoglobin breaks down to release oxygen (see Fig. 6.16). The oxygen diffuses

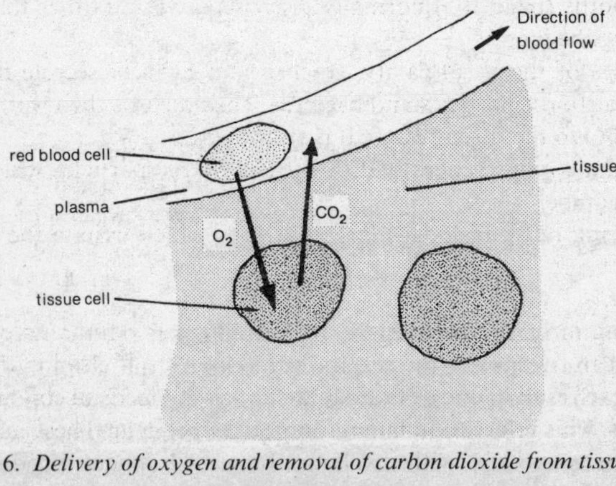

Fig. 6.16. *Delivery of oxygen and removal of carbon dioxide from tissues*

through the capillary wall, across the tissue fluid, and into respiring cells.

At the same time, carbon dioxide diffuses in the other direction and, on reaching the blood, combines with water to form bicarbonate ions. The blood, now rich in carbon dioxide and low in oxygen (deoxygenated), travels through the circulatory system back to the lungs (see Fig. 7.2).

Remember, oxygen is carried in the blood as oxyhaemoglobin within red blood cells. Haemoglobin picks up and releases oxygen under the following conditions:

$$\text{Haemoglobin} + \text{oxygen} \underset{\substack{\text{low } [O_2] \text{ – at body tissues} \\ \text{high } [CO_2]}}{\overset{\substack{\text{high } [O_2] \text{ – at the lungs} \\ \text{low } [CO_2]}}{\rightleftharpoons}} \text{Oxyhaemoglobin}$$

Carbon dioxide is carried in the blood as bicarbonate which is formed and broken under the following conditions:

$$\text{Carbon dioxide} + \text{water} \underset{\substack{\text{high } [O_2] \text{ – at the lungs} \\ \text{low } [CO_2]}}{\overset{\substack{\text{low } [O_2] \text{ – at body tissues} \\ \text{high } [CO_2]}}{\rightleftharpoons}} \text{Bicarbonate}$$

The enzyme **carbonic anhydrase**, found in red blood cells, catalyses the above reaction.

6.6. *Respiratory Disorders*

Several mechanisms in the body ensure that the respiratory surface in the lungs is kept clean and that the respiratory passages remain unblocked:

1. The epiglottis (page 105) normally prevents anything other than air from entering the lungs.
2. The linings of the nasal cavity, trachea and bronchi secrete mucus which traps airborne particles and bacteria. The mucus is then wafted by cilia (page 106) to the throat where it is swallowed.
3. The coughing reflex generally dislodges any large particles which do enter the windpipe.
4. The sneezing reflex dislodges large particles which irritate the nasal lining.

The breathing in of smoke, dust particles and other pollutants over a long period can overpower the respiratory system's self-cleaning mechanisms and can result in one or more of the following medical conditions:

1. **Bronchitis**. This refers to inflammation of the bronchi. These air passages become narrowed with swelling. Increased mucus production

causes them to become partially blocked. Coughing and soreness result and may become extremely serious (chronic) bronchitis. Bronchitis is more common among smokers and those living in smoky industrial towns. The passing of the Clean Air Act in 1964 has reduced the incidence of bronchitis in industrial towns.

2. **Respiratory infections** such as colds or more seriously, pneumonia have a higher incidence among those who live in smoky areas. Smoke particles slow down the action of cilia and cause mucus to build up in the respiratory passages and lungs. As a result, microbes and dust particles accumulate, increasing the risk of infections and damage to tissues.

3. **Pneumoconiosis** is the name given to a group of diseases caused by inhaling various forms of dust. These conditions are often associated with a particular occupation. **Silicosis**, for example, is usually associated with mining or quarrying. Particles of silica cause damage to the lungs, and stimulate increased production of mucus. This may lead to coughs, breathlessness, bronchitis, and an increase in respiratory infections. **Asbestosis**, caused by inhaling asbestos fibres, has similar symptoms, and can also lead to lung cancer. Blue asbestos, strongly linked with causing cancer, has now been banned in the United Kingdom.

In industry, regulations have been brought in to reduce the likelihood of contracting these dust-related diseases. The wearing of masks, ventilation of factories, and the damping down of dusty surfaces, are some of the preventative measures used.

4. **Allergic reactions**. Coughing, sneezing and a runny nose (caused by excess mucus production) are the symptoms of hay fever – a condition triggered by the pollen of plants. These symptoms arise because some people are **allergic** to the proteins and waxes that make up the outer coats of pollen grains; their bodies treat these chemicals as 'foreign' and produce antibodies against them (see page 132).

5. **Cigarette smoke**, like all forms of smoke, consists of fine particles suspended in a hot gas. Cigarette smoke is particularly dangerous to health because:

(a) Particles in tobacco smoke slow down the beating action of cilia; in heavy smokers the cilia may become gradually destroyed. The characteristic 'smoker's cough' is an attempt to clear the mucus which accumulates when the cilia do not operate effectively. The reduction in air-cleaning efficiency means that particles and microbes are more likely to enter the lungs and cause physical damage and respiratory infection. In extreme cases this can lead to irreparable damage and loss of alveoli – a condition called **emphysema** – whereby the victim becomes short of breath.

(b) Certain tars and other chemicals in tobacco smoke are **carcinogenic** (cancer-promoting) and over time may cause **lung cancer**.
(c) **Nicotine** in tobacco smoke is absorbed into the blood and has a number of effects on the body. Nicotine causes an increase in blood pressure both by constricting (narrowing) arteries and increasing the rate of heartbeat. Nicotine also raises blood fatty acid concentrations and causes platelets to stick more readily together and to the walls of arteries. The net result is to increase the likelihood of arterial deposition (arteriosclerosis) – 'hardening of the arteries' (page 127). This in turn increases the likelihood of a 'stroke' or a heart attack (see page 127).
(d) **Carbon monoxide** in tobacco smoke diffuses into the blood and combines with haemoglobin to form carboxyhaemoglobin. This prevents some of thc haemoglobin from combining with oxygen and thus reduces the blood's oxygen-carrying capacity (see also page 117).
(e) **Tar and other chemicals** deposited in the lungs by smoking line the alveoli and impair gas exchange. This reduces a person's ability to undertake physical exercise.
(f) **Smoking during pregnancy** can result in reduced physical development of the foetus, a lower birth weight and a greater likelihood of still birth or infant death. Nicotine in the mother's blood is delivered to the baby via the placenta and seems to reduce tissue growth directly. Carbon monoxide also reaches the baby's blood, reduces its oxygen-carrying capacity and deprives the foetus of much needed oxygen.

6.7. *Gas Exchange in Flowering Plants*

The main gas exchange surface in flowering plants is the surface of the **spongy mesophyll cells** in the leaf. Gases enter from, and escape to, the atmosphere through the **stomatal pores** (see Fig. 6.17 and also Fig. 5.2, page 50).

Gas exchange in green plants is complicated by the process of photo-

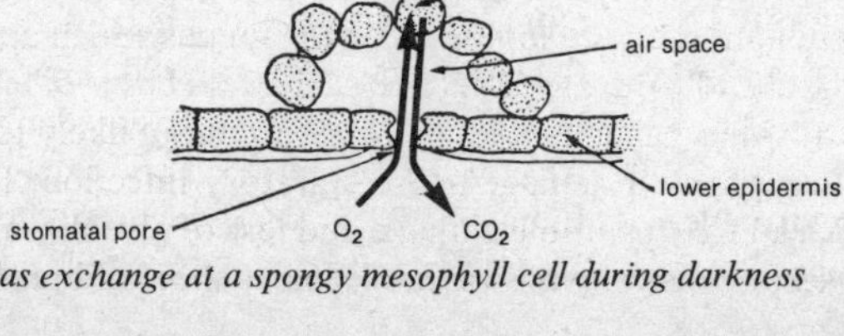

Fig. 6.17. *Gas exchange at a spongy mesophyll cell during darkness*

synthesis, which in many ways is the reverse of respiration. The overall exchange of gases varies with light intensity (see page 62).

During daylight hours, the rate of photosynthesis is usually much greater than respiration, so overall the mesophyll cells take up carbon dioxide and give off oxygen. This is the reverse of the direction of gas movement at the respiratory surface of animals.

During darkness, plants respire but do not photosynthesize, and so there is an overall release of carbon dioxide and uptake of oxygen. This is the same result as that at the respiratory surface of animals.

In woody plants (page 309) gas exchange also occurs at the **lenticels** of the stem (Fig. 6.18). These small openings in the bark of woody stems are filled with loosely packed cells separated by air spaces.

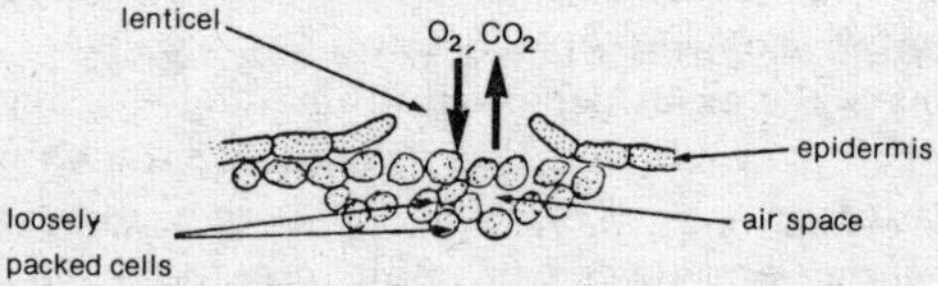

Fig. 6.18. *Gas exchange at a lenticel in the stem of a woody plant*

Definitions

Respiration
Gas exchange
Breathing (ventilation)

Key Words

Respiratory or gas exchange surface	Spiracles	Bronchioles
Mitochondria	Tracheal system	Diaphragm
Anaerobic	Gill	Intercostal muscles
Aerobic	Gill filaments	Inspiration
Alcohol (ethanol)	Gill arch	Expiration
Lactic acid	Gill lamellae	Tidal air
Lime water	Pharynx	Vital capacity
Soda lime	Operculum	Residual air
Potassium hydroxide	Alveoli	Total capacity
Haemoglobin	Lungs	Bicarbonate
	Thoracic cavity	Carbonic anhydrase

Epidermis	Pleural cavity	Spongy mesophyll cells
Tracheoles	Pleural membranes	Stomatal pores
Tracheae	Trachea (windpipe)	Lenticels
Sodium hydrogencarbonate	Bronchi	

Questions

1. (a) Describe an experiment which demonstrates that germinating seeds give off heat.
(b) What are the differences between aerobic and anaerobic respiration?
(c) Explain how the products of anaerobic respiration are used in
(i) making alcoholic drinks
(ii) making bread.

2. The table below shows the relationship between heart disease, age and smoking.

	Death rate per 100,000 from heart disease		
Cigarettes daily	*Aged under 45*	*Aged 45–54*	*Aged 55–64*
0	*5*	*120*	*445*
1–14	*45*	*220*	*750*
15–24	*55*	*390*	*840*
25+	*110*	*405*	*1030*

(a) Draw a histogram of these results.
(b) What is the relationship between smoking and heart disease?
(c) Name two other ways in which smoking damages health.

3. Name precisely *the gas exchange surface for the following animals*
(i) a mammal
(ii) a fish
(iii) an insect
(iv) Name three *features shared by* all *the above animals' gas exchange surfaces and* one *feature shared by* only two *of them (state which two share this feature).*

7. *Transport*

Every living cell in an organism needs to take up substances, such as food and oxygen, and has to get rid of waste. These substances enter and leave cells mainly by diffusion (Section 4.1).

In single-celled organisms, such as *Amoeba* (page 315), *Chlamydomonas* (page 297) and *Spirogyra* (page 299), diffusion through the body surface and across the cell membrane supplies the needs of the organism. This also applies to the simpler multicellular organisms which have a thin body wall.

In larger multicellular organisms, however, cells on the inside are too far away from the external environment to obtain supplies and get rid of wastes simply by diffusion. Thus, special transport systems have developed to deliver and take away substances. A transport system links the cells inside the body to the external environment via specialized structures. In mammals, for example, the lungs exchange respiratory gases, the gut absorbs soluble food, and the kidneys remove wastes. In higher plants, for example, root hairs absorb water and minerals from the soil. Such structures perform their functions on behalf of all cells in the body.

In certain invertebrate groups, e.g. the annelids (page 317) and arthropods (page 319), and in all vertebrates (page 323), the transport system consists of a fluid (blood) pumped through vessels by a muscular pump (the heart). This is called a **circulatory system**.

In flowering plants (page 308), there are two transporting fluids; the phloem and xylem. Both flow through systems of tubes (see page 134).

Detailed knowledge and understanding of the transport systems of both mammals and higher plants are required.

The components of the mammalian circulatory system are:

1. circulatory fluids (blood and lymph)
2. a system of vessels to carry the fluids (arteries, veins, capillaries and lymph vessels)
3. a pump (the heart).

7.1. *The Mammalian Circulatory Fluid: Blood*

Components of blood

Blood is a tissue (page 24) made up of three parts:
1. a liquid part, **plasma**, making up 55 per cent of blood
2. **blood cells** of several types, making up 45 per cent of blood
3. small cell fragments called **platelets**.

1. **Plasma** is a straw-coloured solution, 90 per cent water, and contains the following dissolved substances:
(a) mineral salts, e.g. sodium chloride ($NaCl$) and sodium bicarbonate ($NaHCO_3$)
(b) food substances – monosaccharides, glucose, amino acids, fatty acids, glycerol and vitamins
(c) plasma proteins – fibrinogen and antibodies
(d) hormones (chemical messengers)
(e) nitrogenous wastes, mainly urea.

The relative amounts of these substances vary from one part of the body to another (page 128) and from one minute to the next. For example, the glucose level in plasma increases after a meal and the level of the hormone adrenalin is increased by emotional stress (page 177).

2. **Blood cells**: there are two main types:
(a) red blood cells (erythrocytes)
(b) white blood cells (leucocytes).

The features of these cells are summarized in Table 7.1.

Red blood cells are minute biconcave discs. They have no nucleus and hence have a short life span (about 3 months). They are broken down by the liver (page 81) and are constantly replaced by cells in the bone marrow. Red blood cells contain two important substances: **haemoglobin**, which combines with oxygen for transport (page 117), and **carbonic anhydrase**, which converts carbon dioxide to bicarbonate for transport (page 117).

White blood cells are larger than red blood cells, have a nucleus and are present in much fewer numbers. There are three main types and each plays a vital role in the body's defence against disease (page 131).

3. **Platelets** (thrombocytes): these are very small cell fragments formed by cells in the bone marrow. At the site of an injury, platelets, together with damaged cells, release an enzyme **thrombokinase** which initiates blood clotting. A sequence of events follows:

Table 7.1. The structures in mammalian blood

	Appearance	Number/mm^3	Site of Formation	Function
Red blood cells (erythrocytes) 	i) biconcave (thin central region) ii) no nucleus iii) red due to presence of haemoglobin	5 million	Red bone marrow	Oxygen transport (see text) Bicarbonate formation (see text)
White blood cells (leucocytes) Two main types:				
1. Granulocytes 	i) granular (grainy) cytoplasm ii) lobed (lumpy) nucleus	7,000	Red bone marrow	Phagocytic – engulf bacteria
2. Agranulocytes Two types: a) Lymphocytes 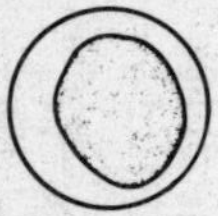	i) non-granular cytoplasm ii) large nucleus	2,500	Lymph nodes	Secrete antibodies
b) Monocytes 	i) non-granular cytoplasm ii) kidney-shaped nucleus	500	Red bone marrow	Phagocytic – engulf bacteria
Platelets 	i) cell fragments	¼ million	Red bone marrow	Blood clotting

Note: In a sample of normal human blood the above structures are present in the ratio red blood cell : white blood cell : platelet
500 : 1 : 25

thrombokinase
↓
prothrombin ——————————→ *thrombin*
(inactive enzyme present in blood plasma) — in the presence of calcium ions — (active enzyme)
↓
fibrinogen ——————————→ *fibrin*
(soluble protein in plasma) — (insoluble) forms a meshwork of fibres at the site of injury
↓
blood cells trapped in meshwork seal up leak and prevent entry of disease organisms
↓
meshwork containing trapped cells dries to form a protective scab and allows healing of the wound underneath

This sequence of events is complex, but is necessary to make sure clotting does not take place by accident within blood vessels. A clot blocking the inside of a blood vessel stops the flow of blood. If this occurs in the brain it can cause a 'stroke', or in the heart muscle, a coronary thrombosis (heart attack).

Treatment of bleeding

It is important to reduce the amount of blood lost until the clot is formed. When people cut themselves the following procedure should be followed:

1. Press on the wound with your fingers. Where possible the wound should be covered with a piece of clean cloth first.
2. Lay the person down and, if possible, raise the wound to above the level of the heart.
3. If bleeding is severe or the person is in shock, call an ambulance.

NOTE. A cut caused by a blow to the head should not be pressed on with the fingers in case there is a fracture underneath it. A dressing should be applied to the wound and bandaged in place. The patient should be sitting in an upright position while waiting for the ambulance.

Functions of blood

1. **Formation of tissue fluid**: every cell in the body of a mammal lives in a liquid environment called tissue fluid, which is derived from blood. Tissue fluid is basically blood plasma that has been forced out of blood capillaries (see page 128).

Blood has several major transport functions:

2. **It transports oxygen** from the lungs to respiring tissues in the form of oxyhaemoglobin within red blood cells (page 108).
3. It **transports carbon dioxide** from respiring tissues to the lungs in the form of bicarbonate (page 108).
4. It **transports the products of digestion** (glucose, amino acids, fatty acids, glycerol) from the small intestine to the liver, and then to the rest of the body.
5. It **transports nitrogenous wastes** (e.g. urea from the liver) from tissues to the kidneys.
6. It **transports hormones** from the glands which secrete them (endocrine glands) to the target tissues on which they act.

Blood **distributes heat** and is involved in **regulating body temperature**:

7. The heat generated by active organs such as the liver and muscles is distributed around the body by the blood.
8. By varying the amount of blood which flows to the skin, heat loss from the body can be controlled. In this way, body temperature can be regulated (page 159).

The blood plays a major role in **protecting the body against pathogens** (disease-causing organisms):

9. Two types of white blood cell, granulocytes and monocytes (page 118) engulf pathogens by phagocytosis.
10. Another type of white blood cell, the lymphocytes (page 118), secrete chemicals called antibodies which attack pathogens (page 132).
11. The plasma protein fibrinogen initiates blood clotting at the site of an injury (page 119). Clotting helps prevent invasion of the body tissues by pathogens.

Notice 'Functions of blood' has eleven points grouped under four sections. With proper explanation and attention to detail you should be able to give clear and full answers.

7.2. *Blood Vessels*

Muscular contractions of the heart pump blood into **arteries**, which divide inside body tissues and organs to form first arterioles, and then very fine vessels called **capillaries**. These join together to form first venules and then **veins**, which carry the blood back to the heart.

At the arterial end of capillaries, plasma is forced into the surrounding tissues to form **tissue fluid**. Most of this fluid returns to the blood at the venous end of capillaries, but some enters **lymph vessels** to form **lymph** (page 129). The lymph travels through the lymphatic system and empties into the bloodstream just before the right side of the heart.

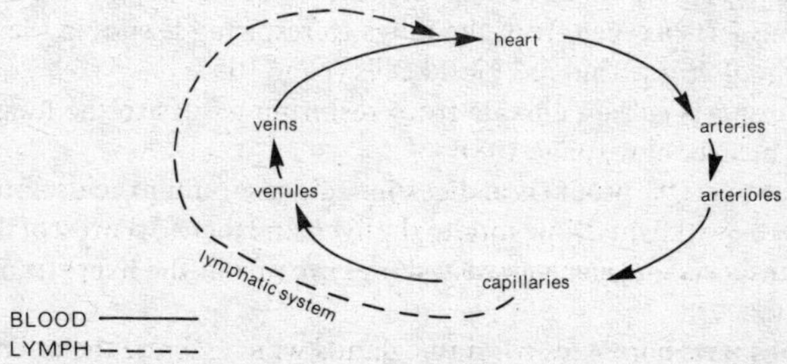

The structure and functions of the different types of blood vessel are summarized in Table 7.2. Notice how structure is related to function.

Arteries have thick walls to withstand high blood pressure and their elastic walls convert the irregular blood flow from the heart into a steady flow to the tissues. Elastic recoil of the arteries gives rise to the pulse which can be felt in the neck and wrist.

The capillaries are small in diameter and very 'leaky'. All the blood's functions depend on the permeability of capillaries, since it is here that substances are exchanged with surrounding tissues.

Blood in veins is at low pressure and its flow towards the heart is assisted by one-way valves and by the action of contracting skeletal muscles (page 197) which squeeze on the veins.

Mammals have a **double circulatory system** consisting of two sets of blood vessels connected to the heart. The **pulmonary circulation** carries blood to and from the lungs, while the **systemic circulation** conveys blood to and from tissues throughout the body (see Fig. 7.1).

Having two circulations keeps oxygenated blood separate from deoxygenated blood. Also, it ensures blood is delivered to tissues at high pressure. After passing through one set of capillaries the blood is returned to the heart to have its pressure raised.

Table 7.2. Mammalian blood vessels: their structure and function

	Structure	Function
Arteries 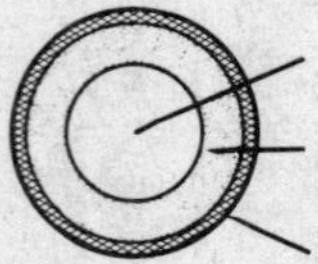 T.S. artery	i) thick-walled with small lumen (bore) ii) thick muscle layer with many elastic fibres iii) no valves	i) carry blood away from the heart under high pressure ii) carry oxygenated blood (except pulmonary artery) iii) elastic recoil helps reinforce pumping action of the heart and maintain steady blood flow iv) dilation (widening) or constriction (narrowing) of arteries diverts blood flow to particular regions of the body
Veins 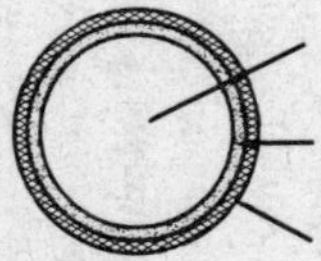 T.S. vein 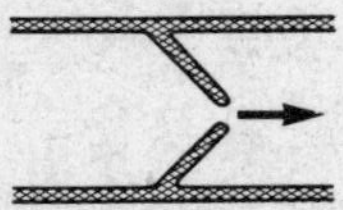 L.S. vein showing valve	i) thin-walled with large lumen ii) thin muscle layer with few elastic fibres iii) valves present	i) carry blood towards the heart under low pressure ii) carry deoxygenated blood (except pulmonary vein)
Capillaries 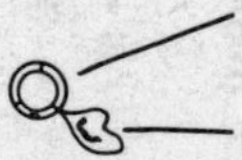	i) wall one cell thick ii) no muscle layer iii) no valves	i) carry blood between arteries and veins ii) plasma forced out of capillaries under high pressure to form tissue fluid (see page 128) iii) granulocytes and monocytes can squeeze out of capillaries and into surrounding tissues

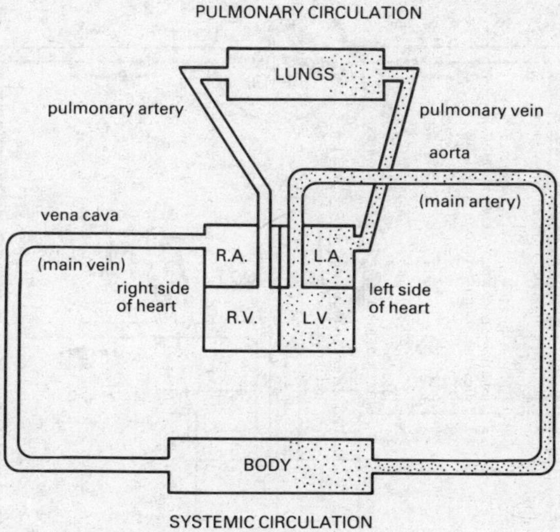

Fig. 7.1. *The double circulation of mammals (oxygenated blood shown stippled)*
R.A. = Right atrium *L.A. = Left atrium*
R.V. = Right ventricle *L.V. = Left ventricle*

The major blood vessels and their connections with the heart are shown in Fig. 7.2. The following points should help you remember their names:

1. **Arteries** carry blood away from the heart and to organs.
2. **Veins** (except portal veins) carry blood away from organs and to the heart.
3. **Pulmonary** refers to the lungs.
4. **Hepatic** refers to the liver.
5. **Renal** refers to the kidneys.

Questions may be asked about the route blood takes from one organ to another. For example:

Q. What is the correct sequence of blood vessels through which blood travels from the left ventricle to the kidneys and back to the heart in a mammal?

(*a*) *Dorsal aorta, renal artery, renal vein, posterior vena cava.*
(*b*) *Dorsal aorta, hepatic artery, hepatic portal vein, posterior vena cava.*
(*c*) *Anterior vena cava, renal vein, hepatic artery, dorsal aorta.*
(*d*) *Anterior vena cava, hepatic vein, renal artery, pulmonary vein.*
(*e*) *Hepatic portal vein, hepatic vein, dorsal aorta, renal artery.*

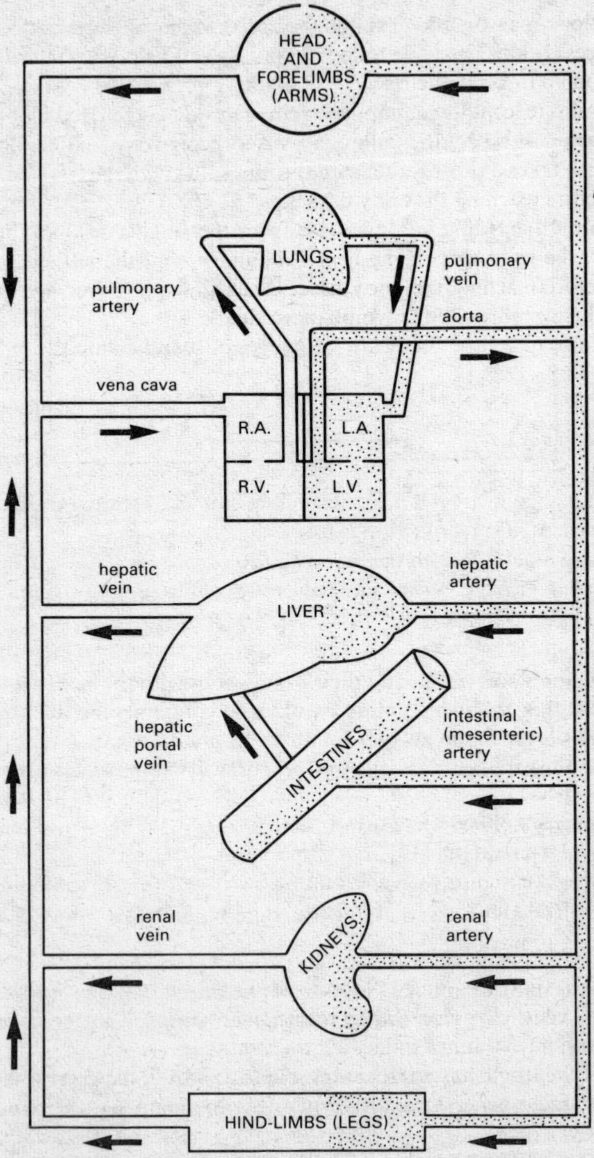

Fig. 7.2. *Connections between the heart and main blood vessels in man (arrows show direction of blood flow; oxygenated blood shown stippled)*

Questions may be asked about the route taken by certain dissolved substances. In this case, the following points will help you (refer to Fig. 7.2 as you go through them):

1. The **veins** (except the pulmonary vein) carry deoxygenated blood.
2. The **arteries** (except the pulmonary artery) carry oxygenated blood.
3. **Oxygen** takes a short route from the lungs, through the left side of the heart, before reaching the body tissues.
4. **Dissolved foods** take a long route from the small intestine, through the liver, the right side of the heart, the lungs, and the left side of the heart, before reaching the body tissues (most fats, however, bypass the liver and travel through the lymphatic system).

Refer also to section 7.4 below, 'Changes in Blood Composition'.

7.3. *The Heart*

The mammalian heart is a muscular pump situated between the lungs in the thoracic cavity. Its function is to:

1. circulate blood through the blood vessels
2. generate a blood pressure sufficient to force plasma out of capillaries (page 128).

The mammalian heart is actually two pumps joined together, one on the left and one on the right. Both pumps have two chambers, an **atrium** (auricle) and a **ventricle**. Valves in the heart ensure that blood flows through the heart in one direction only.

An accurate, fully labelled drawing of the heart should be learnt – you may be expected to draw or label one. Before doing this, it is worth looking at a simplified diagram of the heart showing the basic features (see Fig. 7.3 overleaf).

Compare this figure with the one below it showing the mammalian heart in ventral view.

Notice the following features:

1. The right ventricle pumps blood to the lungs.
2. The left ventricle pumps blood round the body.
3. The atria have thinner walls than the ventricles.
4. The left ventricle has a thicker muscular wall than the right ventricle, because it must generate a force sufficient to pump blood round the entire body.
5. One-way valves are positioned at the exits of the heart chambers.

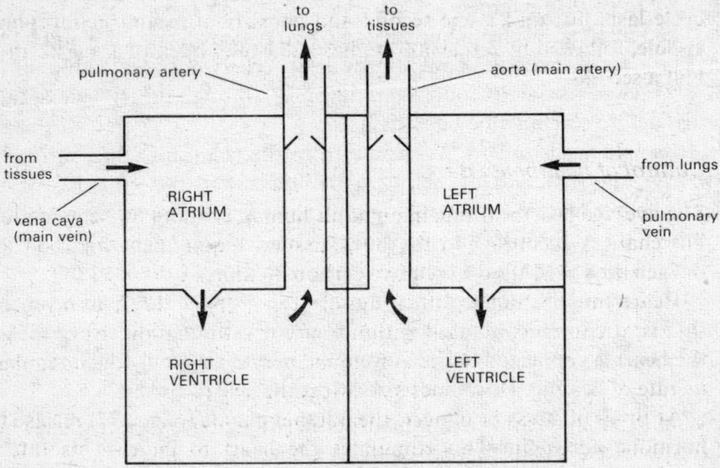

Fig. 7.3. *The mammalian heart (diagrammatic): ventral (front) view*

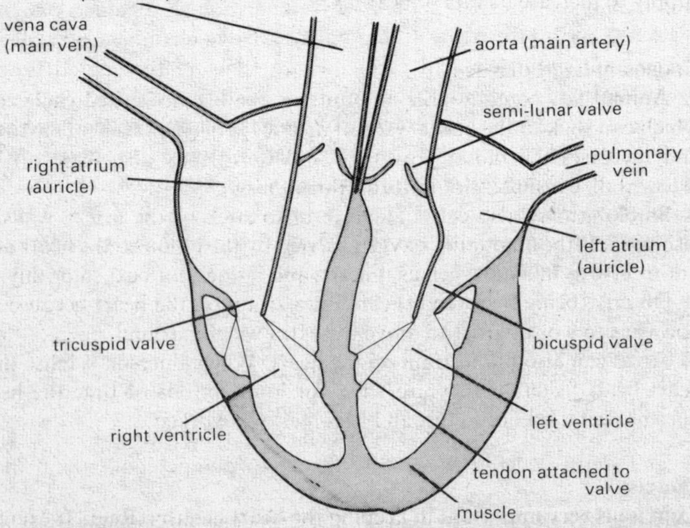

Fig. 7.4. *Vertical section through the mammalian heart: ventral (front) view*

Cardiac cycle

The pumping action of the heart is described by the cardiac cycle. The cycle lasts just under one second and consists of a contraction phase, **systole**, followed by a relaxation phase, **diastole**. Each phase lasts about half a second.

Control of heartbeat rate

The average heartbeat rate in an adult human is about 70 beats/minute, but changes according to the body tissues' requirement for food and oxygen, and their need to remove carbon dioxide.

Heart muscle, unlike other muscle types (page 197), is **myogenic**, that is, it contracts by itself without nervous stimulation. Nevertheless, the heart is connected to the autonomic nervous system which regulates its rate of beating. Hormones also affect the heartbeat rate.

At times of stress or danger, the adrenal glands (page 177) release the hormone adrenalin. This stimulates the heart to increase its rate of beating to prepare the body for 'fight or flight' (page 177).

When the level of carbon dioxide in the blood rises, for example during exercise, then the heart is stimulated by the autonomic nerve supply to increase its rate of beating.

Causes of heart disease

1. Animal fats contain large amounts of a substance called **cholesterol** which can stick to the inside walls of arteries. This makes them narrower and increases the blood pressure. Cholesterol may also flake off the artery wall, causing a clot to form (thrombosis).
2. Smoking may also cause cholesterol to stick to the artery walls. It also reduces the amount of oxygen carried by the blood so the heart does not receive as much oxygen as it needs and it does not beat smoothly.
3. Obesity (being overweight) can put a strain on the heart because the body has to work harder to move the extra weight around.
4. Stress can also put a strain on the heart. When a person is tense their heart beats faster. If they are tense for long periods of time the heart does not get a chance to rest and becomes overworked.

Exercise

Exercise is very important in keeping the heart healthy. Regular exercise can make the heart muscle stronger in the same way that other muscles can be strengthened by training. During exercise the blood is pumped round the body faster than usual and this helps to keep the blood vessels clear.

It is very dangerous for an unfit person suddenly to exercise vigorously; they must build up slowly to get the heart used to the extra demands made upon it.

7.4. *Changes in Blood Composition*

As blood passes round the body, its composition changes depending on the function(s) of the organ or tissue through which it is flowing. For example, when blood flows through any tissue it loses oxygen and dissolved foods, and gains carbon dioxide. The overall changes in blood composition can be summarized as follows (see table opposite):

The table is by no means complete but does show the major changes. Hormones (page 176) have not been included.

7.5. *Formation of Tissue Fluid*

All the cells in a mammal's body are no more than 0.025mm from a capillary – the source of the tissue fluid which surrounds them. All the cells' food and oxygen are obtained from the tissue fluid, and all the cells' wastes are lost into it.

Tissue fluid is formed at the arterial end of capillaries where the high blood pressure forces fluid out through the thin walls of the capillary and into spaces between the cells (Fig. 7.5). It is similar to plasma in containing oxygen and dissolved foods but does not contain protein molecules because they are too large to pass through capillary walls. Red blood cells, platelets and the non-phagocytic white blood cells, lymphocytes, are similarly unable to pass through the capillary walls.

Phagocytic white blood cells, the granulocytes and monocytes (page 118), can squeeze out of capillaries and into tissues.

At the venous end of capillaries the blood pressure is much reduced, and the blood (now very thick because of the plasma it has lost) absorbs back tissue fluid by osmosis (see page 40). Some of the tissue fluid filters into small blind-ended lymph vessels to become **lymph**.

Region of the body	*Blood gains*	*Blood loses*
All tissues	Carbon dioxide Nitrogenous waste	Oxygen Food (monosaccharides, amino acids, etc.)
Lungs	Oxygen	Carbon dioxide Water
Small intestine	Dissolved foods – water, minerals, vitamins, monosaccharides, amino acids, glycerol, fatty acids	
Liver	Controlled amounts of glucose and fats Urea Heat	Glucose (for storage as glycogen) Fatty acids and glycerol Excess amino acids Old red blood cells
Kidneys		Urea Water Salts
Bones	Red blood cells Granulocytes and monocytes (types of white blood cell)	Iron (for Hb in red blood cells) Calcium and phosphate (for bone growth)
Skin	Vitamin D	Heat Salts and urea (in sweat)
At the veins of the neck where the lymphatic system empties into the bloodstream	Lymph containing fats and lymphocytes	

7.6. *The Lymphatic System*

The lymphatic system is a very extensive system of vessels which drains off tissue fluid from capillary beds all over the body. The lymph so formed is similar in composition to tissue fluid but in addition contains lymphocytes and certain proteins. Where the lacteals (page 76) empty into the lymphatic system, the lymph contains a large amount of fat.

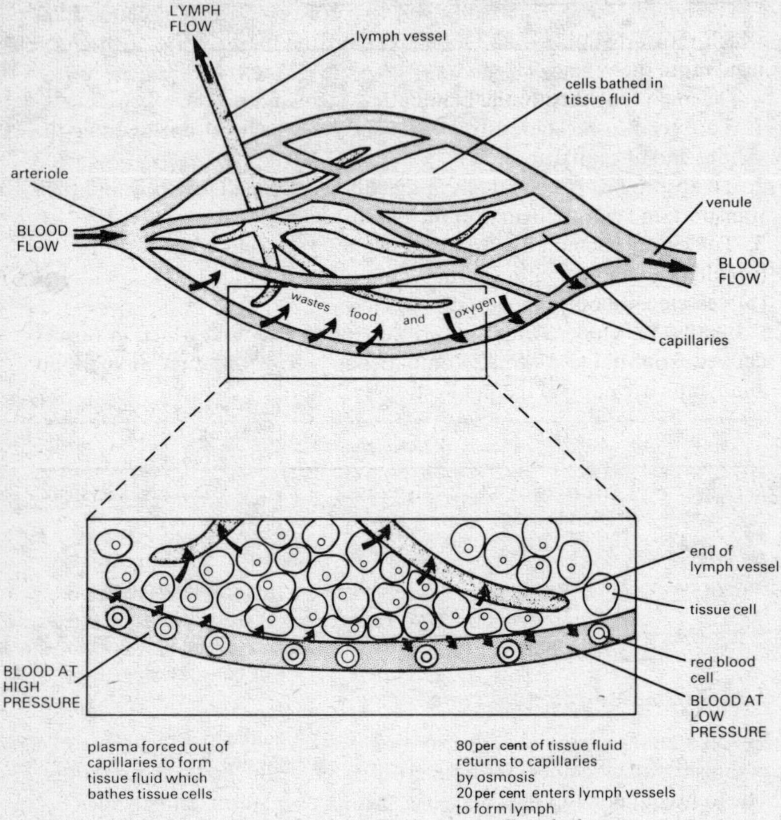

Fig. 7.5. *Formation of tissue fluid and lymph at a capillary bed*

At intervals along the lymph vessels are collections of cells forming **lymph nodes**. These manufacture lymphocytes and antibodies (page 132) and contain phagocytic cells which filter out, ingest and destroy bacteria. Lymph nodes in the armpits, groin and neck are grouped together to form lymph glands. The tonsils and spleen are essentially large lymph glands. Lymph nodes near the site of an infection often become swollen and painful as a result of the increased activity of the white blood cells they contain.

The flow of lymph in lymph vessels is maintained by the action of skeletal muscles pressing on the vessels and by the presence of one-way valves. In this respect, lymph vessels are similar to veins (page 122).

The lymph vessels eventually join to form two large lymphatic ducts which return lymph to the bloodstream at the major veins of the neck just before the vena cava.

The main functions of the lymphatic system are:

1. To return to the bloodstream substances which do not re-enter the venous end of capillaries.
2. To absorb fats from lacteals in the villi of the small intestine and then transport and empty them into the bloodstream.
3. To destroy bacteria by producing white blood cells which:

(a) filter out and engulf bacteria

(b) secrete antibodies (page 132).

Finally, remember lymph is derived from tissue fluid which in turn is derived from blood plasma. Comparisons between plasma, tissue fluid and lymph are popular exam questions.

	Blood plasma	*Tissue fluid*	*Lymph*
Location	Heart and blood vessels	Tissues	Lymphatic system
Differences in contents	Plasma proteins Fat droplets Red blood cells	No plasma proteins No fat droplets No red blood cells	Some proteins Fat droplets No red blood cells

7.7. *Defence Against Disease*

The mammalian body is under continual bombardment by microscopic organisms which can cause disease if they enter the body and multiply. The four commonest types of disease-causing organism (**pathogen**) are:

1. certain bacteria (page 294)
2. viruses (page 292)
3. certain protozoa (page 315)
4. certain fungi (page 302).

The majority of infectious human diseases are caused by bacteria or viruses.

To combat these organisms the body has a sophisticated system of defence. The first line of defence attempts to prevent organisms entering the body, and the second attempts to kill or immobilize them if they do enter.

First line of defence

Physical barriers and active agents normally prevent pathogens from entering the body.

Physical barriers:

1. The skin epidermis (page 157) is impermeable.
2. Healthy mucous membranes in the mouth, nose, windpipe and oesophagus are impermeable.
3. Hairs lining the nasal cavity filter out airborne particles and pathogens.

Active agents:

1. On the skin, the oily substance sebum produced by sebaceous glands (page 157) contains an anti-bacterial agent.
2. On the skin, sweat produced by sweat glands (page 157) contains an anti-fungal agent.
3. In the eyes, tears contain an anti-bacterial agent.
4. In the stomach, the acid-containing gastric juice kills bacteria.
5. In the nose and windpipe, cilia (page 106) trap airborne particles and waft them towards the mouth where they are swallowed.
6. If there is surface damage to the skin, blood clotting helps prevent entry of pathogens.

Second line of defence

If the first line of defence fails, and pathogens enter the body, the second line of defence is called into play:

1. Phagocytic cells in the lymph nodes, and in the blood (granulocytes and monocytes), engulf and destroy bacteria.
2. Interferon, a protein secreted by individual cells throughout the body, can stop the multiplication of viruses.
3. The immune system (see below) produces antibodies which attack pathogens.

Immunity

Definition: Immunity *is the body's ability to resist infection by the use of antibodies.*

Immunity to a particular disease may be acquired naturally or may be induced artificially.

The basis of immunity lies in the body's ability to manufacture **antibodies** which can attack **antigens**. Antigens are chemicals on the surface of, or secreted by, pathogens. Antibodies are made and secreted by lymphocytes (page 118).

Immunity can be classified as active or passive.

Active immunity is when an organism achieves immunity to a particular disease by making its own antibodies to combat the disease organism.

Passive immunity is when an organism gains immunity by receiving antibodies 'ready-made' from another organism.

Active immunity can be acquired naturally, by recovering from the disease, or induced artificially, by vaccination.

Examples of naturally acquired active immunity are resistance to measles, mumps or chicken pox, after recovering from the illness.

Alternatively, active immunity can be induced artificially by vaccination. **Vaccination** is the injection into the body of disease antigens which trigger the body's immune system to produce appropriate antibodies. Antigens of the disease organism can be given in the form of:

(a) Toxoids, which are toxins (poisons) produced by the pathogen and then inactivated (made harmless) before injection. The diphtheria vaccine is an example.

(b) Killed vaccines, which contain the dead pathogen. The poliomyelitis (polio) vaccine is an example.

(c) Attenuated (weakened) living vaccines, which contain a weakened strain of the pathogen. An example is the BCG vaccine against tuberculosis (TB).

Passive immunity can occur naturally or be induced artificially.

An example of naturally acquired passive immunity is the protection against infection a foetus gains from antibodies which are supplied by the mother (page 229).

A tetanus injection is an example of artificially induced passive immunity. After a surgical operation a person may be given an anti-tetanus vaccination. The vaccine contains blood serum from a horse which has been infected with the tetanus bacterium. (NOTE. Serum is plasma with the fibrinogen removed, so that it does not clot.) The horse has manufactured the appropriate anti-tetanus antibodies, and these are being given ready-made to the patient. The vaccine gives the patient short-term protection against tetanus infection.

Rejection

The body's ability to recognize foreign tissues causes problems when organs are transplanted. The transplanted organ is attacked by the body's defence systems and as a result may be **rejected**. Transplant patients are given drugs to help the body accept the new organ by suppressing their immune system. This means that the patient is very susceptible to infections and has to be kept in sterile conditions for some time after the transplant operation.

Hay fever is also a form of rejection. Particles such as pollen grains come into contact with the sensitive tissues lining the respiratory system. The body's defensive responses then produce the symptoms of hay fever.

7.8. *Transport in Plants: Xylem and Phloem*

In flowering plants (page 308) there are two transport systems: **xylem** and **phloem**. Xylem is the fluid found inside xylem vessels. It transports water and minerals from the roots to the stem and the leaves. Phloem in phloem sieve tubes transports food substances from the sites where they are made, notably the leaves, to the growing parts and storage regions of the plant.

Together, the xylem vessels and phloem sieve tubes form the tube-like **vascular bundles** which run through roots and stems and can be seen as veins in the leaf (Fig. 7.6). Each vascular bundle contains a number of xylem vessels and phloem sieve tubes (Fig. 7.7). In the root and stem of a dicot, the sieve tubes are to the outside of the xylem vessels (Fig. 7.6 and 7.7).

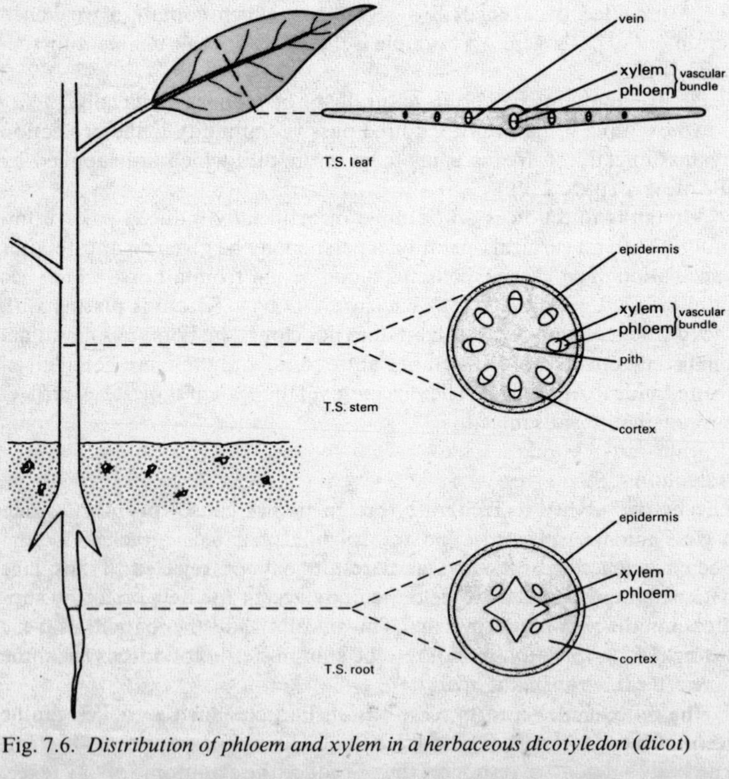

Fig. 7.6. *Distribution of phloem and xylem in a herbaceous dicotyledon* (*dicot*)

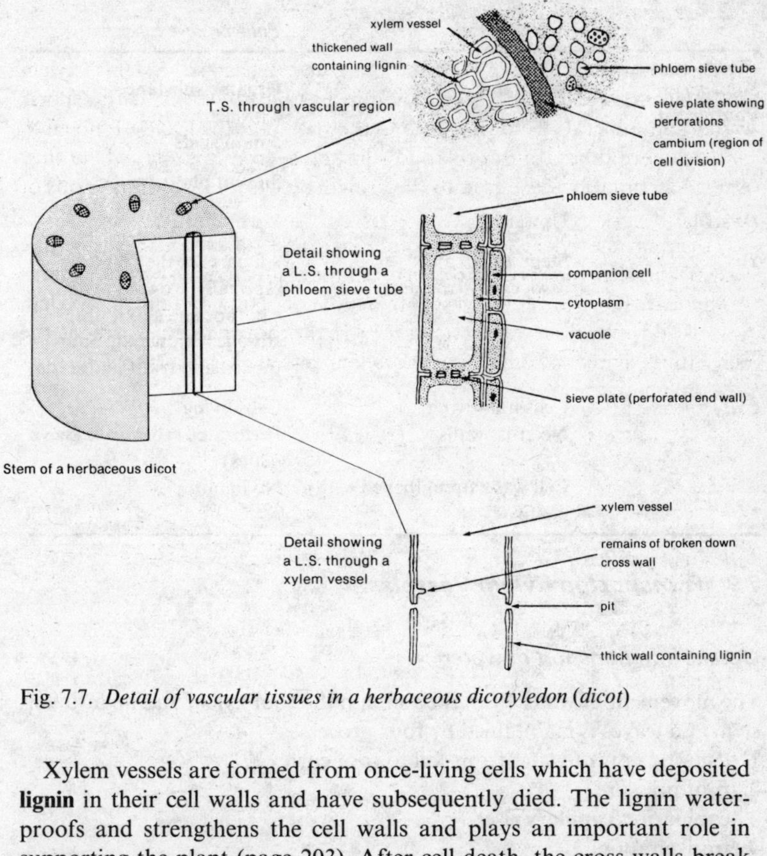

Fig. 7.7. *Detail of vascular tissues in a herbaceous dicotyledon* (*dicot*)

Xylem vessels are formed from once-living cells which have deposited **lignin** in their cell walls and have subsequently died. The lignin waterproofs and strengthens the cell walls and plays an important role in supporting the plant (page 203). After cell death, the cross walls break down (Fig. 7.7) and the string of cells running from root to leaf forms a continuous tube. At intervals, holes or pits enable substances to be transported sideways out of the vessel (page 144).

The phloem sieve tubes are made up of living cells. One cell is connected to the next by a perforated end wall called a sieve plate (Fig. 7.7). The cells have no nucleus and their cell walls do not contain lignin. Companion cells are closely associated with the sieve tube cells and are thought to be actively involved in the transport of substances within the tube.

The differences between xylem vessels and phloem sieve tubes can be summarized in a table:

	Xylem vessels	*Phloem sieve tubes*
Substances transported	Water and mineral salts	Organic substances, e.g. sucrose and amino acids
From	Roots	Sites of photosynthesis
Direction	Upwards	Up or down
To	Stem, leaves and associated structures	All parts of the plant, especially growing regions, e.g. root and shoot tips; storage regions, e.g. bulbs or tubers; and fruits and seeds
Cell structure	Cells non-living No cross-walls	Cells living Perforated cross walls (sieve plates)
	Cell walls strengthened with lignin	No lignin

7.9. *Transport in Xylem Vessels*

Uptake and transport of water

The movement of water from the soil into the root xylem and up into the stem and leaves is maintained by four processes:

1. osmotic flow of water from soil to root cortex
2. root pressure
3. capillarity (capillary rise)
4. transpiration.

1. **Osmotic flow** from soil to root cortex (steps 1 and 2 in Fig. 7.8). Water is taken up (absorbed) from the soil and into the plant root by osmosis. This occurs mainly through the root hairs of epidermal cells. Root hairs are found in a narrow zone 0.5 to 1 cm behind the root tip (page 312). They increase the surface area of the root for absorption of water, and several million are found in a single plant.

Soil water is usually found as a thin film around individual soil particles. Root hair cells normally have a higher O.P. (osmotic potential) than the soil water and thus take up water by osmosis. Cells towards the centre of the cortex have a relatively high O.P. and water flows by osmosis from one cell to the next towards the root xylem.

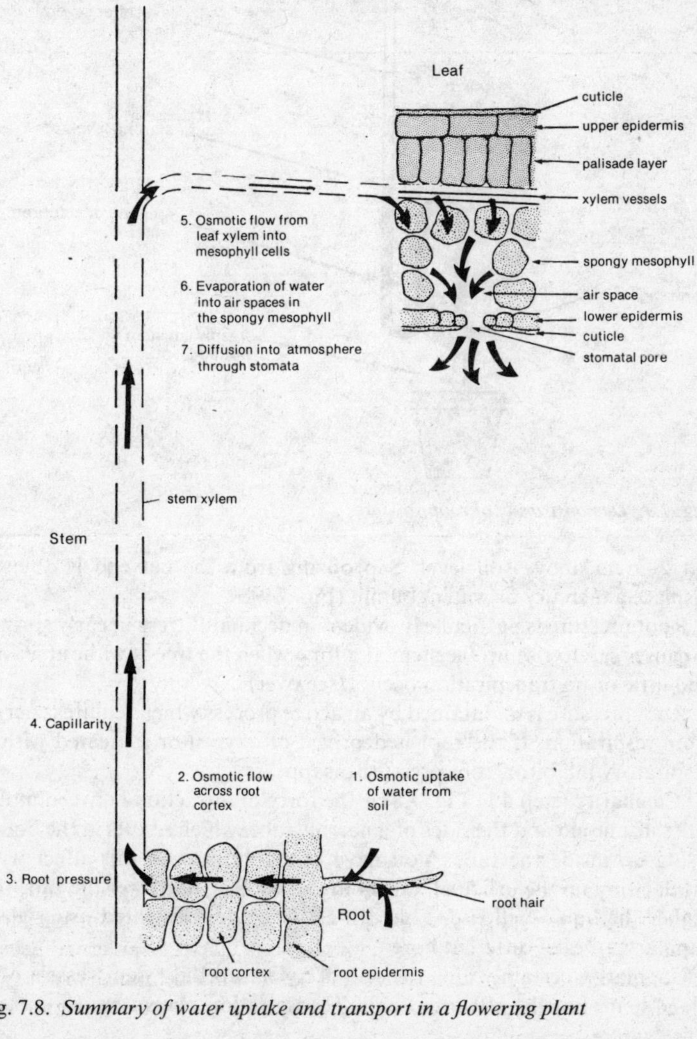

Fig. 7.8. *Summary of water uptake and transport in a flowering plant*

2. **Root pressure** (step 3 in Fig. 7.8): it is not entirely clear how water passes from the root cortex into xylem vessels. It is believed an active transport process (page 44) pumps mineral ions into the root xylem and water then follows by osmosis. This flow of water into the root xylem accounts for root pressure, which can be demonstrated in a plant stem

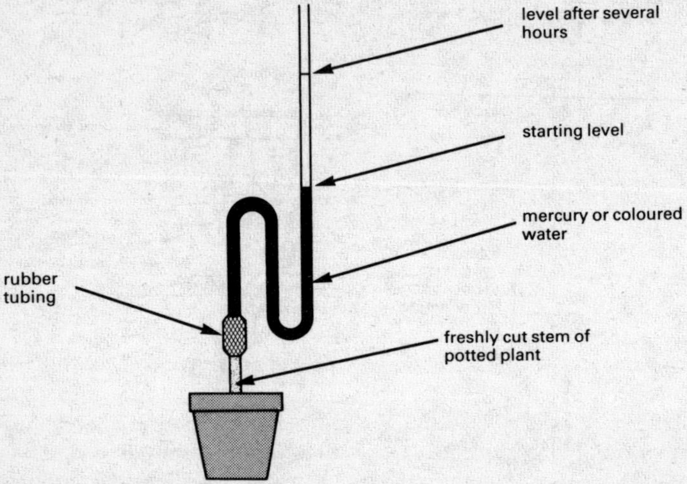

Fig. 7.9. *Demonstration of root pressure*

cut 2–3 cm above soil level. Sap oozing from the cut end is able to displace a mercury or water column (Fig. 7.9).

Root pressure is particularly evident in deciduous trees in early spring. It causes sap to rise up the stem at a time when the tree is without leaves and little or no transpiration occurs (see over).

Root pressure is maintained by an active process which requires energy from respiration. If the root is deprived of oxygen or is treated with a respiratory inhibitor, root pressure disappears.

3. **Capillarity** (step 4 in Fig. 7.8) is the force of attraction between molecules in a liquid and the sides of a narrow tube, which results in the liquid rising up inside the tube. You have probably noticed this effect with drink rising up the inside of a drinking straw. The narrower the tube the higher the liquid will rise. This can easily be demonstrated using glass capillary tubes of different bore.

Capillarity certainly aids the rise of xylem up the xylem vessels of a plant stem, but the effect is small. Transpiration plays a much greater role.

4. **Transpiration** (steps 5–7 in Fig. 7.8) is the loss of water from the plant by evaporation. This evaporation occurs mainly from the surface of spongy mesophyll cells in the leaves, the water being lost mainly through stomata (page 113). A small amount of water is lost through the waxy cuticle of the leaf and through the cuticle and lenticels (page 113) of the stem.

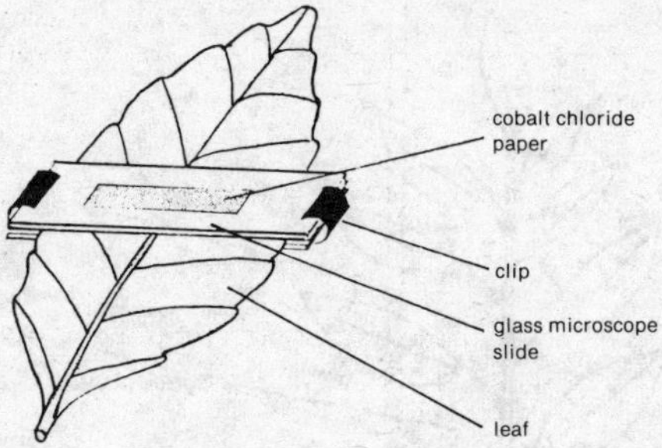

Fig. 7.10. *Using cobalt chloride paper to demonstrate water loss from a leaf*

Definition: Transpiration *is the evaporation of water from the aerial (above-ground) surfaces of the plant.*

Transpiration can be demonstrated by clamping cobalt chloride paper on to a leaf (Fig. 7.10). The paper is blue when dry but turns pink when moist. On the leaf, the paper usually turns pink after an hour or so, indicating that water is being lost from the leaf surface. If paper is clamped to both upper and lower surfaces, which paper would you expect to turn pink first? (see page 49).

During transpiration, the spongy mesophyll cells lose water by evaporation and their O.P. rises. Water is drawn by osmosis from adjacent cells and then in turn from xylem vessels in the leaf. This creates a tension or pull on the xylem in the leaf and stem. This pull is sufficient to draw water up the stem in a continuous flow, called the **transpiration stream**.

The flow of water up the xylem vessels can be demonstrated by placing the cut end of a leafy shoot in water stained with a dye such as eosin red or methylene blue. After several hours, transverse sections of the stem are cut and examined under a microscope. It is seen that only xylem tissue contains the dye, indicating that it is xylem which conducts water up the stem.

Transpiration is a loss of valuable water, and if the water lost by evaporation is not replaced by water uptake, then the plant will wilt. Evaporation from a plant is an unfortunate side effect of having leaves which are designed to allow gas exchange to occur. A leaf structure which encourages gas exchange also allows rapid water loss (see page 113). However, transpiration may have certain useful effects. In hot conditions, it cools the leaf slightly (see also sweating in mammals on page 158). Also, it is argued, by producing the transpiration stream, it draws mineral salts up the plant.

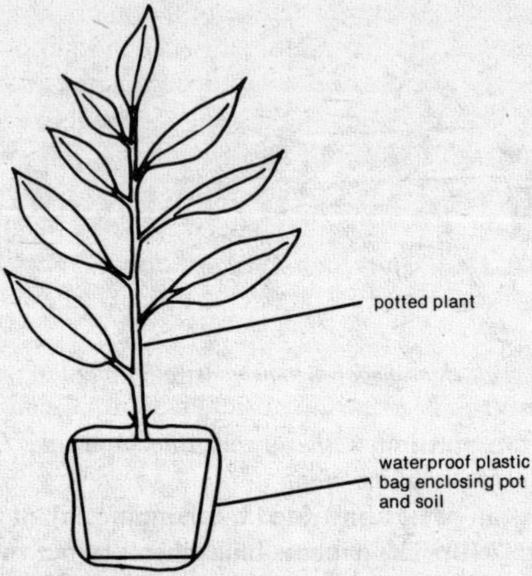

Fig. 7.11. *Apparatus for measuring water loss in a plant*

Rates of transpiration

The rate of transpiration in a plant depends on the water diffusion gradient (page 38) between the air spaces in the leaf, and the atmosphere just outside the stomatal pore. Any factor which increases this gradient will increase transpiration.

The main factors governing transpiration rate are:

1. Temperature: a rise in temperature increases the rate of transpiration by speeding up the evaporation and diffusion of water.
2. Air humidity: if the air is moist (humidity high) the water diffusion gradient will be low and the transpiration rate low. The opposite applies if the air is dry. Dry air favours transpiration.
3. Air movement (wind) will carry away water molecules that have diffused through stomata. Strong air movements (wind) stop water molecules accumulating outside the stomatal pore. This helps maintain a steep diffusion gradient and thus increases transpiration.
4. Any factor which causes stomata to close will effectively stop transpiration. Light and water availability control opening and closure of stomata:

(a) In general, stomata open during daytime and close at night.
(b) Stomata close at times of severe water shortage, when wilting occurs.

In summary, transpiration rates are increased by a rise in temperature, a lowering in air humidity, or an increase in air movement. Transpiration rates are decreased by a drop in temperature, a rise in humidity, or a decrease in air movement. In addition, absence of light or a shortage of water causes stomata to close and thus stops transpiration.

The rate of transpiration under different conditions can be measured using a simple apparatus such as that shown in Fig. 7.11. A well-watered potted plant is used and the pot and soil enclosed in a waterproof plastic bag to reduce evaporation from the soil. The plant is weighed, left under experimental conditions for a set number of hours, and then reweighed. Any weight loss is presumed to be due to loss of water by evaporation from the stem and leaves, i.e. transpiration.

The experiment can be repeated under different sets of conditions. For example, the plant can be placed in normal light, and then darkness, and the transpiration rates compared. By placing the plant in a transparent enclosure, and introducing varying amounts of water, the plant can be exposed to different humidities. A hair drier (with the heating element switched off) can be used to generate varying wind conditions. In each case, the experiment should be repeated under the same laboratory conditions, with only one factor altered – light, temperature, humidity, or wind.

Typical results under laboratory conditions are:

Condition	*Rate of transpiration as measured by weight loss*
Normal light	High
Darkness	Low
High humidity	Low
Low humidity	High
High wind	High
Low wind	Low
High temperature	High
Low temperature	Low

The amount of water lost per hour by transpiration is small compared to the overall weight of plant and pot. Using the above apparatus, experiments should be carried out over several hours or days to get a measurable weight loss.

Rates of transpiration can be measured more rapidly using a **potometer** such as that shown in Fig. 7.12. This apparatus measures water uptake

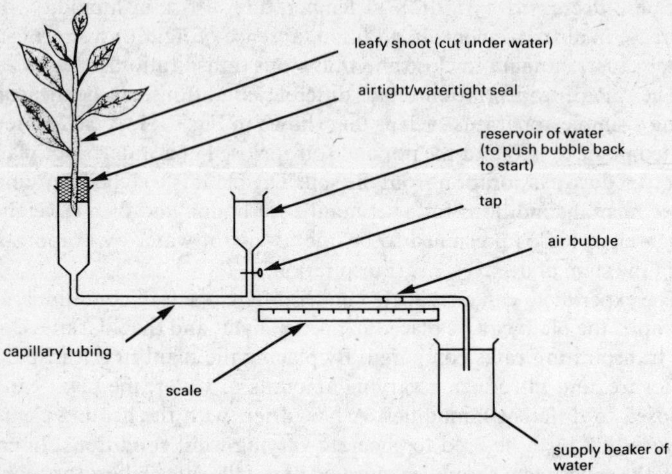

Fig. 7.12. *A potometer, used for measuring water uptake by a plant shoot*

rather than water loss. Water uptake and loss are not exactly equal, i.e. not all the water taken up by the plant is lost by evaporation. Some water, but only about 5 per cent, is used by the plant in photosynthesis and in forming new tissue. Transpiration rates measured using a potometer are very slightly overestimated, but the difference is negligible.

Certain precautions are necessary when using a potometer. Refer to Fig. 7.12 as you go through them.

1. The stem of a leafy shoot is used. It must be cut under water and kept immersed while being connected to the apparatus. This prevents air from entering the xylem and causing an 'air lock'.
2. The apparatus is set up as shown and then left under experimental conditions for 15 minutes, to reach equilibrium, before a bubble is introduced into the capillary tubing.
3. The rate of travel of the bubble (in cm/hour) is used to measure the rate of water uptake. If the bore of the tube is known, the rate of water uptake can be calculated in cm^3/hour.

Using such an apparatus it is found that the environmental conditions that affect transpiration rate usually affect water uptake in the same way. Water uptake is increased by high temperature, low humidity and moving air, and decreased by low temperature, high humidity and still air. In the absence of light, water uptake decreases. These results suggest that transpiration is the major process controlling water uptake by the

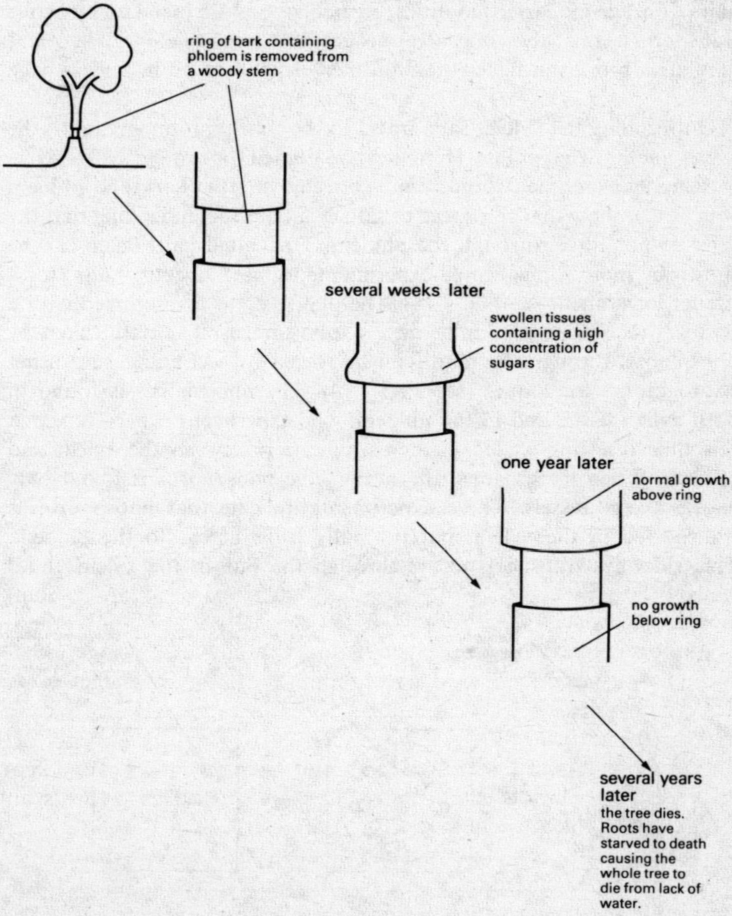

Fig. 7.13. *The ringing (girdling) of a tree*

stem. This is, in fact, the case. Transpiration is the major factor governing water movement through the plant.

Uptake and transport of mineral salts

Dissolved mineral salts are absorbed into the plant as **ions**. The ions are taken up by diffusion in the osmotic flow of water through the root hairs. If needed in large quantities, particular ions are taken up by active transport, against the concentration gradient (see page 44). The plant may absorb one ion faster than another, depending on its needs (page 64).

On reaching the xylem, salts travel in the transpiration stream to the leaves and growing points. Here they are needed for protein synthesis.

'Ringing experiments' on stems, i.e. removing the bark and phloem (Fig. 7.13), show that transport of salts is unaffected, indicating that the salts do not travel through the phloem. The actual path taken can be shown by more sophisticated experiments using radioactive salts (for a similar experiment see Fig. 7.14). The plant's roots are supplied with a radioactive isotope of a mineral, e.g. phosphorus. The path taken by phosphorus through the plant can be traced by sectioning, and using photographic techniques (see Fig. 7.14). The phosphorus is found in both xylem tissue and in the phloem. The experiment can be repeated this time inserting waxed paper as a barrier between the xylem and phloem. When this is done, the radioactive phosphorus is found only in the xylem vessels. These experiments indicate that phosphorus is transported in the xylem, but normally leaks across to the phloem. This sideways transport occurs through the pits in the xylem vessel walls.

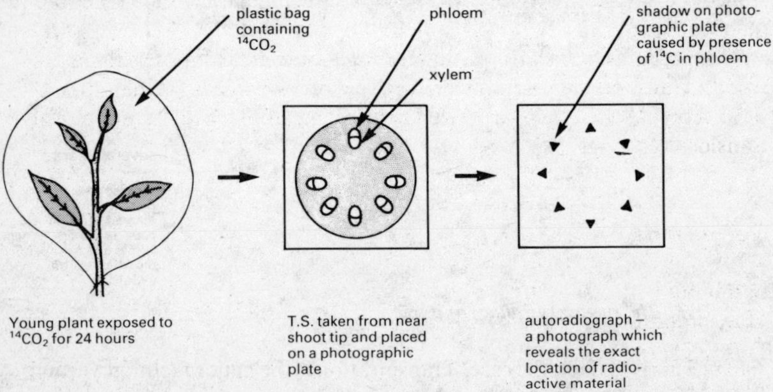

Fig. 7.14. *An experiment which demonstrates that organic compounds are transported to the stem phloem*

7.10. ***Transport in Phloem Sieve Tubes (Translocation)***

Translocation is the term usually used to describe the transport of organic substances from sites where they are made to other parts of the plant. The sugars and amino acids made in the leaf after photosynthesis (page 48), are transported in the phloem to growing parts and storage regions of the plant.

Definition: Translocation *is the transport of soluble organic substances within the plant.*

Movement of organic substances can occur up the stem to developing leaves, flowers and fruits, or down the stem into growing regions or storage organs in the root or stem.

The transport of organic substances in the phloem sieve tubes can be demonstrated in two sets of experiments, Figs 7.13 and 7.14. If a tree is 'ringed' by removing a ring of phloem-containing bark, then the tree will eventually die. Organic substances made in the leaves and transported down the stem cannot reach the roots and the roots eventually die.

NOTE. The flow of xylem from the roots up the stem is unaffected.

A second experiment involves the use of carbon dioxide labelled with radioactive carbon (^{14}C). The ^{14}C is incorporated into organic substances during photosynthesis and the path then taken by the ^{14}C in the plant can be traced.

The mechanism of translocation through the phloem is not well understood. Transport depends on the activity of living cells in the phloem sieve tubes. If these cells are killed or their metabolic activity inhibited, translocation stops.

Definitions

Immunity
Transpiration
Translocation

Key Words

Circulatory system	Monocytes	Lymph nodes
Blood	Thrombokinase	Pathogens
Lymph	Prothrombin	Antibodies

Arteries
Veins
Capillaries
Lymph vessels
Heart
Plasma
Platelets (thrombocytes)
Red blood cells (erythrocytes)
White blood cells (leucocytes)
Granulocytes
Lymphocytes

Thrombin
Fibrinogen
Fibrin
Tissue fluid
Double circulatory system
Pulmonary circulation
Systemic circulation
Atria (auricles)
Ventricles
Systole
Diastole
Myogenic

Antigens
Active immunity
Passive immunity
Xylem
Phloem
Xylem vessels
Phloem sieve tubes
Vascular bundles
Lignin
Stomata
Root pressure
Capillarity
Potometer
Ions

Questions

1. (a) Draw a table to show the differences between arteries and veins.
(b) Describe, in detail, the route taken by blood from the time it enters the heart from the vena cava to the time it leaves the heart in the aorta.

2. Use a large labelled diagram to show how tissue fluid is formed.

3. (a) Explain how evaporation of water from the leaves of a plant helps to draw water up the xylem from the roots.
(b) Name two other factors involved in the movement of water through the xylem.
(c) How do the following environmental factors affect the movement of water in plants?
(i) humidity
(ii) temperature
(iii) air movements.

4. (a) Give two functions of the lymph system.
(b) Name three places in the body where lymph nodes (glands) are found.
(c) If you cut your foot and it goes septic you get a pain in your groin. Explain.

8. *Keeping Conditions in the Body Constant (Homeostasis)*

8.1. *Homeostasis*

Definition: Homeostasis *is the maintenance of a constant internal environment.*

In practice, when dealing with a mammal, homeostasis means keeping the **blood** constant in composition and temperature. Remember, it is blood which gives rise to the tissue fluid which bathes the cells in the body (see page 128).

The advantage of homeostasis is that it makes the functioning of the body independent of changes in the external environment. The cells can be kept at their optimum conditions and will function most efficiently. Any variations in pH or temperature can have a seriously inhibiting effect on enzyme action and will lower metabolism (see page 20).

Homeostasis is an active process and requires energy.

In a mammal, the following features of tissue fluid must be kept constant:

1. the balance of chemical constituents such as salts, dissolved foods and waste products
2. the osmotic potential (O.P.), as determined by the relative amounts of water and dissolved substances (page 42)
3. the level of carbon dioxide
4. temperature.

In general we recognize four major sets of organs (homeostatic organs), which regulate these features.

1. the **liver**, which, for example, regulates blood glucose level (page 81)
2. the **lungs**, which regulate the carbon dioxide level in the blood (page 104).
3. the **kidneys**, which regulate the level of salt and water in the blood, and hence regulate the blood's O.P.; they also control the level of nitrogenous wastes and poisons in the blood.
4. the **skin**, which regulates body temperature by increasing or decreasing heat loss through the body surface.

Two important aspects of homeostasis are **excretion** and **osmoregulation**.

Definition: Excretion *is the removal from the body of the waste products of metabolism.*

Table 8.1. *The excretory products and excretory structures of mammals and flowering plants*

	Mammals	*Flowering plants*
	Mammals excrete gases or dissolved solids	Plants excrete gases or undissolved solids
Excretory products:	Carbon dioxide (from respiration)	Carbon dioxide (from respiration)
	Water (from respiration or excess absorbed)	Water (from respiration)
	Nitrogenous wastes, e.g. urea (from protein metabolism)*	Oxygen (from photosynthesis)
	Bile pigments (from breakdown of red blood cells)	Plants excrete few nitrogenous wastes. They make their own proteins and do not have an excess to break down and remove
	Mineral salts (excess of those absorbed)	
Excretory structures:	Lungs – excrete carbon dioxide and water vapour (see page 108)	Stomata – excrete oxygen, carbon dioxide and water vapour (see page 113)
	Kidneys – excrete water, salts, nitrogenous wastes and certain poisons (see page 151)	Whole leaves – some deciduous trees, e.g. oak, transfer wastes to the leaves which are then shed in autumn
	Skin – excretes salts, water and urea from sweat glands (see page 157)	Heartwood – some solid wastes are deposited in the wood at the centre of tree stems
	Liver – excretes bile pigments into the duodenum	Cell vacuoles – some solid wastes are deposited in the vacuoles of living cells, e.g. silicates in the leaves of grasses

* Animals other than mammals secrete nitrogenous wastes other than urea (see page 151).

Excretion should *not* be confused with egestion (page 65) or secretion (page 65).

Definition: Osmoregulation *is the process by which the osmotic potential (O.P.) of body fluids is kept constant.*

8.2. *Excretion*

The removal of waste products of **metabolism** is a vital part of homeostasis. It prevents the accumulation of substances which are poisonous or would otherwise interfere with the normal functioning of body cells. These excreted waste substances are called **excretory products**.

The term excretion has also come to include the removal from the body of excess or poisonous substances which have been absorbed. In the case of water, for example, it is not possible to tell whether excreted water has come from metabolism (e.g. respiration) or whether it has been absorbed.

The major excretory products and excretory structures in mammals and flowering plants are shown in Table 8.1.

Students commonly confuse excretion with egestion. Mammalian faeces contain mainly *egested* material, that is, substances which have not been digested and have not been absorbed. These substances have never passed across a cell membrane into the body and so cannot be excreted. The faeces contain only a small amount of excreted material such as bile salts.

Insects (page 319) are different in this respect. Their faeces contain a large proportion of excreted material. Their Malpighian tubules (roughly equivalent of kidneys) excrete the nitrogenous waste, uric acid, into the gut. It is removed from the body as part of the faeces.

Unicellular animals and plants such as *Amoeba* (page 315) and *Chlamydomonas* (page 297), and the smaller multicellular animals, do not need special excretory organs. Their surface area is large in relation to their volume and waste products can simply diffuse out of the body as quickly as they are formed.

8.3. *Osmoregulation*

Both plants and animals have the problem of maintaining their body fluids at a relatively constant osmotic potential (page 42). However, animals have two problems which plants do not have:

1. **Animal cells do not have cell walls:** if water enters by osmosis the cells must either be surrounded by liquid of the same O.P., as in multicellular organisms, or the cells must have special mechanisms to remove excess water, as in unicellular organisms.

In **fresh water**, body fluids have a higher O.P. than the surrounding water, and water will flow into the body across any semi-permeable

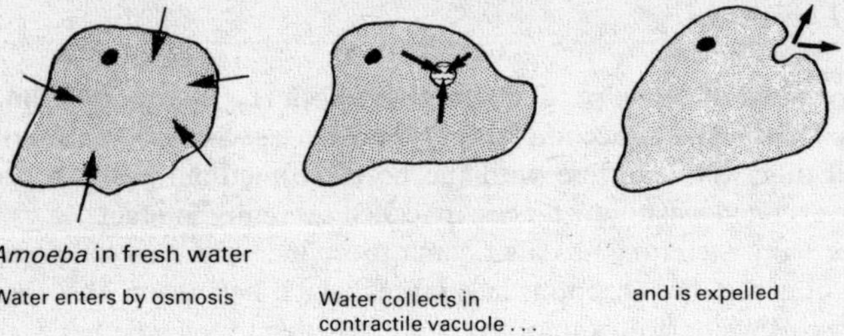

Fig. 8.1. *Osmoregulation in* Amoeba

surface. In *Amoeba* (page 315), excess water flowing in by osmosis is pumped out by a contractile vacuole (Fig. 8.1). This requires energy from respiration. In bony fish (page 323), most of the body surface is impermeable because of its layer of scales and mucus. Water enters by osmosis across the gill epithelium and is also absorbed through the gut wall. Excess water is excreted in urine produced by the kidneys.

In **sea water**, the body fluids of invertebrates (page 291) have about the same O.P. as the surrounding sea water. Hence an *Amoeba* capable of living in both sea and fresh water will have a contractile vacuole in fresh water, but will lose it in sea water. The situation in vertebrates is more complex. The body fluids of bony fish have an O.P. slightly *below* that of sea water. They tend to *lose* water by osmosis. They get round this problem by drinking sea water and pumping out excess salts across their gills. They produce small volumes of concentrated urine.

Terrestrial (land) animals have the added problem of water loss by evaporation.

2. **Animals produce nitrogenous wastes** and water is used in their removal from the body. This can be a source of serious water loss.

Remember, nitrogenous (nitrogen-containing) wastes are derived from the breakdown of unusable proteins and amino acids (see Table 8.2).

$$\text{protein} \rightarrow \text{amino acids} \rightarrow \text{amine } (NH_2) \text{ group} \rightarrow \begin{cases} \text{uric acid} \\ \text{ammonia} \\ \text{urea} \end{cases}$$

With the exception of mammals, animals living in dry environments where there is a serious problem of water loss, tend to excrete their nitrogenous waste in a semi-solid form. Mammals and aquatic animals excrete nitrogenous wastes dissolved in water.

Table 8.2. *Types of nitrogenous waste*

Nitrogenous waste	*Details*	*Animal*
Ammonia	Very soluble and very toxic (poisonous)	Excreted as a dilute solution by most aquatic animals other than mammals, e.g. *Amoeba* and bony fish
Urea	Soluble and less toxic than ammonia	Excreted in urine by mammals
Uric acid	Almost insoluble and non toxic. Normally excreted as a white paste	Excreted by terrestrial animals, e.g. insects and terrestrial vertebrates other than mammals

8.4. *The Mammalian Kidney*

The kidneys are the major organs in a mammal concerned with **excretion** and **osmoregulation**. The kidneys remove the following substances from blood:

1. soluble nitrogenous wastes, mainly urea
2. excess water
3. excess salts
4. excess hydrogen ions (to keep blood pH constant)
5. foreign substances such as drugs, poisons, etc.
6. hormones.

All these functions are excretory, and in addition, 2 and 3 are osmoregulatory.

The above substances are excreted by the kidneys in the **urine**.

Elimination of poisons

As well as dealing with poisonous waste products of metabolism the body has to eliminate poisons that enter the body from outside. Alcohol is broken down by the liver before excretion. Some drugs are not broken down and are excreted unchanged in the urine. The body cannot deal with all poisons, however. Lead, for example, cannot be broken down and accumulates in the body. If it is raised above a certain level, it can affect the nervous system of children.

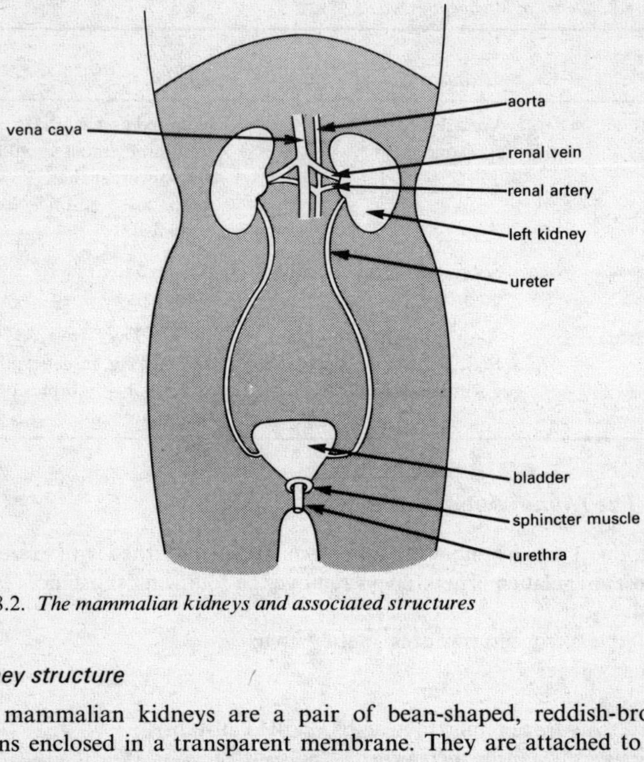

Fig. 8.2. *The mammalian kidneys and associated structures*

Kidney structure

The mammalian kidneys are a pair of bean-shaped, reddish-brown organs enclosed in a transparent membrane. They are attached to the back of the abdominal cavity (Fig. 8.2).

The kidneys are connected to other parts of the body by three vessels:

1. the **renal artery**, which delivers blood to the aorta
2. the **renal vein**, which carries blood away from the kidney to the vena cava
3. the **ureter**, which carries urine away to the bladder.

In vertical section (Fig. 8.3) the kidney consists of a dark outer region, the cortex, and a lighter inner region, the **medulla**. The inner border of the medulla forms several conical **pyramids** which open out into the **pelvis** of the kidney. The pelvis extends into a tube, the **ureter**.

Invisible to the naked eye are the tiny **nephrons** which are the functional units of the kidney. The position of one of these is shown in Fig. 8.3. There are about one million in a human kidney.

Each nephron consists of a bundle of capillaries, the **glomerulus**, surrounded by a **Bowman's capsule**, which empties into a **kidney tubule** (see

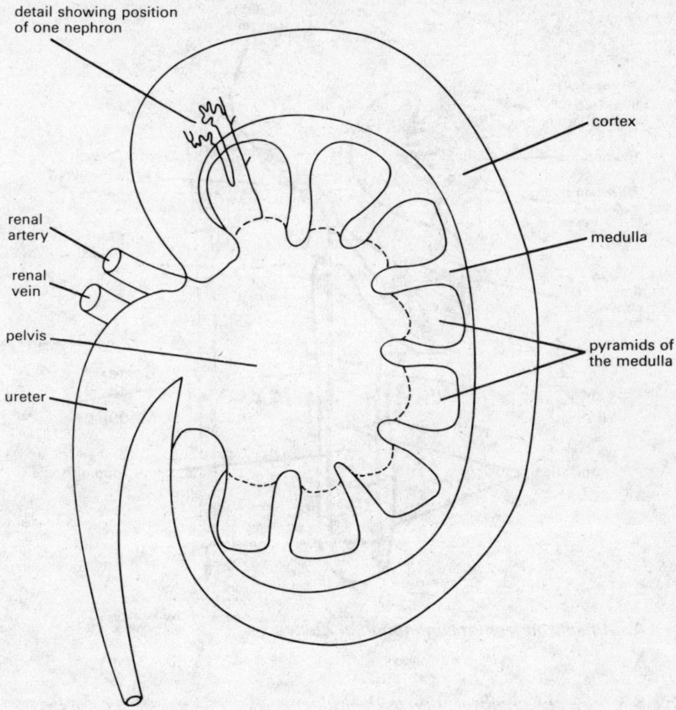

Fig. 8.3. *V.S. of the mammalian kidney*

Fig. 8.4). The tubule is divided into three regions, the **first** (proximal) **convoluted tubule**, the **loop of Henlé** and the **second** (distal) **convoluted tubule**. The tubule leads into a collecting duct which empties into the pelvis of the kidney.

Kidney function

The mammalian kidneys work by filtering the blood, taking back (re-absorbing) from the filtrate those substances which are valuable, and allowing those substances which are harmful or in excess, to leave the body in the urine.

This mechanism involves two processes:

1. **pressure** or **ultra-filtration**
2. **selective reabsorption**.

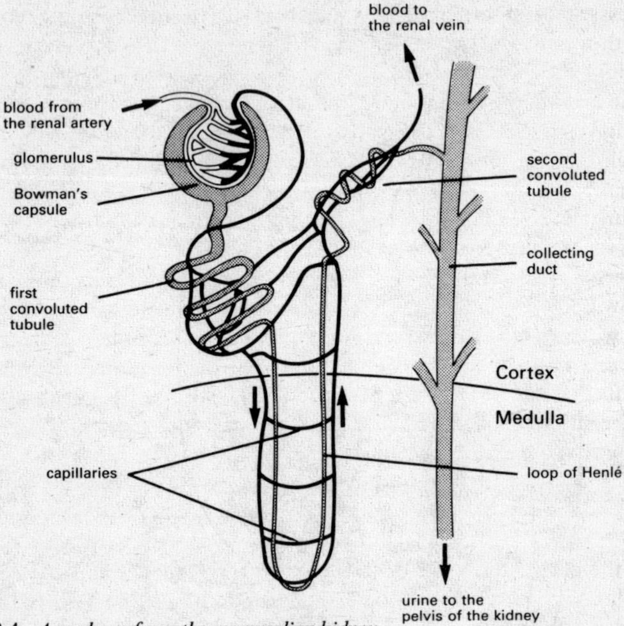

Fig. 8.4. *A nephron from the mammalian kidney*

Table 8.3. *Selective reabsorption in the kidney*

Region	*Substances reabsorbed*	*Mechanism notes*
First convoluted tubule	Glucose, amino acids, salts	Taken up by active transport (page 44) against a concentration gradient. Energy is required
	Water	Taken up by osmosis (page 40). Over 80 per cent of water is reabsorbed here
Loop of Henlé	Water	By osmosis
Second convoluted tubule	Salts	By active transport
	Hydrogen ions	By active transport
	Water	By osmosis *
Collecting duct	Water	By osmosis *

* Uptake of water in the second convoluted tubule and collecting duct is controlled by the hormone ADH according to the body's need to conserve water.

1. **Pressure filtration** (in the Bowman's capsule): blood enters the kidney along the renal artery which branches into many arterioles. One arteriole leads into the glomerulus of each nephon. This arteriole is larger than the one leaving the glomerulus, so that within the glomerulus the blood is at high pressure. This forces plasma and smaller dissolved substances through the capillary walls of the glomerulus and into the cavity of the Bowman's capsule. Larger molecules such as proteins are too large to pass through the capillary walls and they remain in the blood. The fluid within the Bowman's capsule is called the **glomerular filtrate**.
2. **Selective reabsorption** (in the kidney tubule and collecting duct): from the Bowman's capsule the arteriole leading out breaks up into capillaries which wrap around the kidney tubule. This enables valuable substances to be taken back (reabsorbed) into the blood from the glomerular filtrate. The different regions of the kidney tubule reabsorb different substances.

The excretory substances and excess water left in the collecting ducts empty into the pelvis as urine.

Urine

From the pelvis, the urine drips into the ureter which empties into the muscular sac, the bladder, where the urine is stored. Under voluntary control, muscular contractions of the bladder squirt urine into the **urethra** and out of the body. (Do not confuse the ure**thra** with the ure**ter** above.)

A human produces about 1.5 litres of urine per day on average. The urine is about 96 per cent water, 2 per cent urea and 2 per cent salts; it also contains hormones (variously modified) and possibly some poisons or drugs in small amounts. The exact composition of urine varies according to diet, activity and health. For example, people suffering from the condition *diabetes mellitus* lose excess glucose in their urine (glucose is not a normal constituent of urine).

The volume and composition of urine varies according to the body's need to conserve water. In hot weather, the body loses water by sweating (page 159), so to conserve water the kidneys reabsorb as much as possible. As a result, a small volume of concentrated urine is produced.

After drinking large volumes of liquid the O.P. of body fluids may become lowered. In forming urine, the kidneys will reabsorb as little water as possible, and a large volume of dilute urine will be produced.

The amount of water lost in the urine is controlled by the hormone ADH.

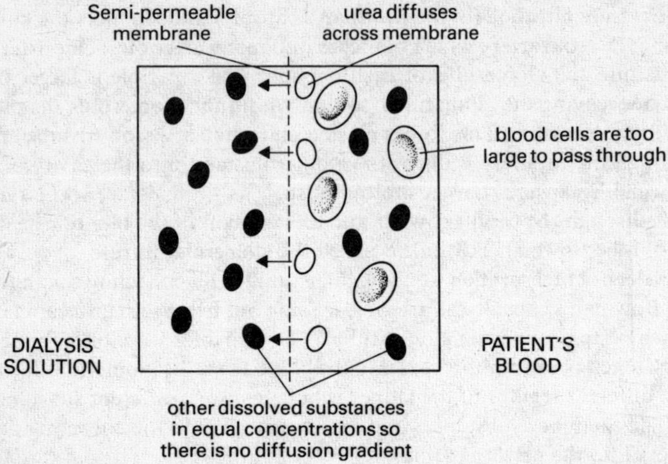

Fig. 8.5. *Dialysis*

Artificial kidneys

People whose kidneys do not work properly can have their blood filtered by a **dialysis machine**. These machines make use of a semi-permeable membrane which allows small molecules to pass across it but large molecules cannot get through.

The patient's blood is pumped into the machine where it is separated by this semi-permeable membrane from a special solution which contains a mixture of salts very similar to that of blood. The main difference is that the solution in the dialysis machine does not contain waste products, such as urea, that are found in blood. Therefore, these waste products diffuse into the dialysis solution down a diffusion gradient. As the solution is continually being renewed, waste products from the patient's blood diffusing into it are removed and disposed of (see Fig. 8.5).

Kidney transplants

When there is severe kidney malfunction, it is possible to transplant a kidney from another person. Tissue from another person is treated as 'foreign' by the body's defence systems and is not always accepted. The new kidney may be attacked and therefore may not function properly. To overcome this problem, transplant patients are given drugs to suppress their immune system. This leaves them open to attack by infections of all kinds and they have to remain in sterile conditions for some time after the operation.

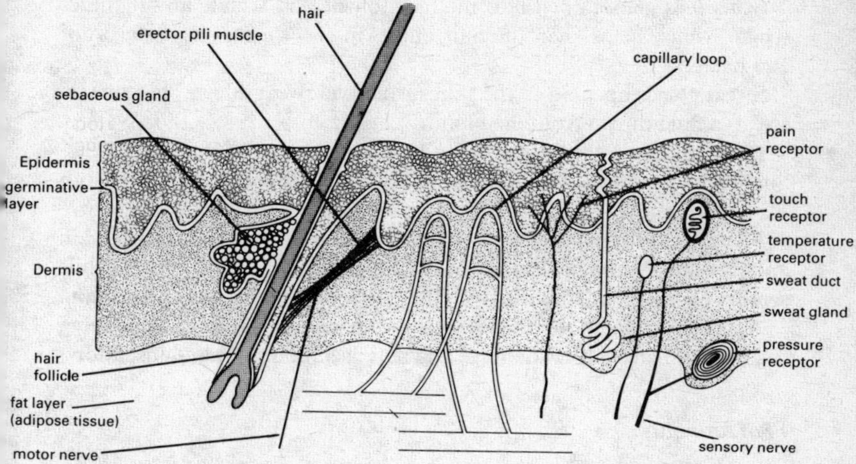

Fig. 8.6. *V.S. of mammalian skin*

8.5. ***The Mammalian Skin***

Skin covers the body surface. It is described as an organ because it is made up of different tissues.

Skin structure

The skin has two main layers:
1. an outer layer, the **epidermis**
2. an inner layer, the **dermis**.

1. **Epidermis:** the epidermis is the protective surface layer many cells thick. Dead cells on top are constantly worn away and replaced by newly formed cells in the **germinative layer**. As cells move towards the surface **keratin** is deposited in them. The cells became flattened, lose their nuclei and die. Keratin waterproofs the surface and reduces water loss from tissues below.

Downfolds of the epidermis form the sweat glands, sebaceous glands and hair follicles, in the dermis.

2. **Dermis:** the dermis contains connective tissue, hair follicles, blood vessels, glands, muscles, nerves and sense organs.

The **hair follicle** secretes the non-living hair. Attached to the hair is an **erector pili muscle** which, when it contracts, pulls the hair upright.

Sebaceous glands open into the hair follicle and secrete an oily fluid, **sebum**, which helps keep the hair and skin surface supple, waterproof and bacteria-free.

Sweat glands open on to the skin surface via sweat ducts. They secrete sweat, a fluid drawn from the blood.

Blood vessels in the dermis bring food and oxygen to the skin tissues. Some give rise to the fluid, sweat, and some form capillary loops. Both play an important part in temperature regulation (section 8.6).

Five types of **sensory receptor** in the skin detect heat, cold, touch, pain and pressure. **Motor nerves** (page 165) stimulate muscles and glands, while **sensory nerves** (page 165) run from the skin sensory receptors to the brain.

A fat layer (adipose tissue) below the dermis protects underlying organs from mechanical damage and acts as a food store and insulator.

Skin function

The skin's functions relate to it being the boundary layer of the body. Its functions are:

1. **Protection:**

(a) The outer dead layer of the epidermis is impermeable and prevents loss of water.

(b) The skin protects against mechanical damage and friction.

(c) A dark pigment **melanin** in the germinative layer prevents u.v. light (from sunlight) penetrating the body and damaging vital organs.

(d) The skin prevents entry of micro-organisms (page 131), the sebum kills bacteria.

(e) The fat layer and hairs reduce heat loss.

2. **Temperature regulation:** sweating, together with widening (dilation) and narrowing (constriction) of blood vessels, controls heat loss through the skin (see page 160).

3. **Vitamin D production:** in the presence of u.v. light (in sunlight) certain fats are converted to vitamin D. Vitamin D is otherwise provided by the diet.

4. **Sensitivity:** the skin is sensitive to touch, heat, cold, pain and pressure. These are protective functions, enabling the body to respond to adverse conditions; they also provide considerable information about the body's surroundings.

5. **Excretion:** water, urea and dissolved salts are excreted in the sweat. The loss of these substances is incidental. The main function of sweat is to increase heat loss.

NOTE. Structures such as teeth (page 82), mammary glands (page 231), nails, hairs, hooves and claws are modified skin tissues.

8.6. *Control of Body Temperature*

Mammals and birds have a constant body temperature and are called **endotherms** (the 'endo-' means that the body temperature is controlled from the inside). Endotherms are commonly referred to as 'warm blooded'. Some textbooks use the term **homoiotherms** to describe these endothermic animals.

Other vertebrates (reptiles, amphibians and fish) and all invertebrates are commonly described as 'cold blooded'. This term is not very accurate because their blood is not always cold but it does vary in temperature quite a lot. The body temperature of these animals is partly determined by the temperature outside their bodies, so they are called **ectotherms**, ecto- meaning the body temperature is related to the external temperature. (**poikilotherms** in some books).

In cold weather the body temperature of an ectotherm will drop and the chemical reactions in its body will slow down. As a result the animal will become sluggish and inactive. An endotherm, on the other hand, is able to keep its body temperature constant during cold weather; this has two major advantages:

1. An endotherm can maintain the optimum temperature for chemical reactions in the body (remember that enzymes are most efficient at a particular temperature).
2. An endotherm can be active all year round and at all times of day.

In humans the body temperature is kept at about 37°C which is the optimum temperature for enzyme action. Death occurs if the body temperature rises much above 42°C or falls much below 30°C. Endothermic organisms usually have a body temperature higher than their surroundings. To maintain this temperature they must generate their own body heat, mainly from respiration. For the temperature to remain constant heat gained must be balanced against heat lost so that:

$$\text{heat gained by body} = \text{heat lost from body}$$

The body gains heat from:

1. metabolism – **respiration** in the liver and muscles releases a large amount of heat energy.
2. heat **radiation** from the sun or nearby hot objects
3. heat **conduction** from the air or any hot objects in contact with the body (including hot food or drink which is ingested).

The body loses heat by:

1. **evaporation** of sweat from the skin or water from the lungs (NOTE. **Latent heat** is needed to convert a liquid into a gas)

2. heat **radiation** to the surroundings
3. heat **conduction** to the surrounding air or any cool objects in contact with the body
4. heat loss by **convection** (warm air next to the body rises and is replaced by colder air)
5. some heat is lost in urine and faeces.

A small structure, the **hypothalamus**, at the base of the fore-brain acts as a thermostat, monitoring blood temperature. If the temperature rises above normal, nerve impulses are relayed to the sweat glands, blood vessels and erector muscles in the skin to increase heat loss. If blood temperature drops below normal, the hypothalamus controls an appropriate set of responses to reduce heat loss and increase heat production. The responses to overheating and overcooling are shown below.

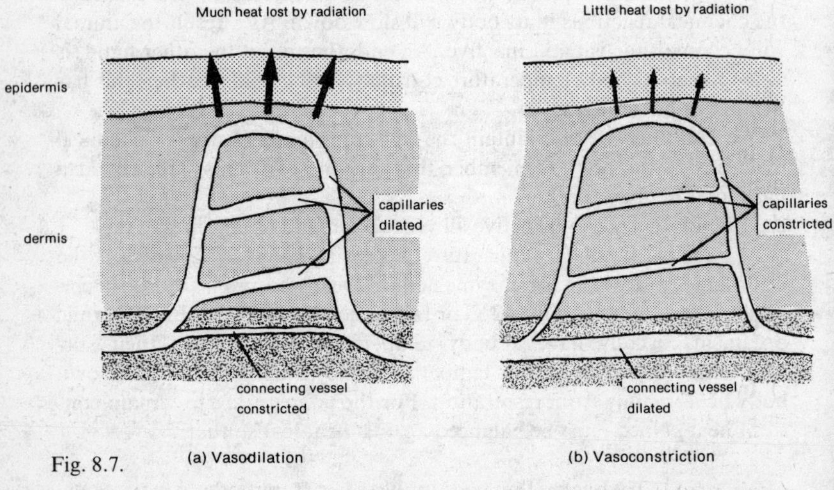

Fig. 8.7.

1. **Control of overheating** (blood temperature in humans rises above 37°C): body temperature may rise during vigorous exercise, when the body is fighting infection, or if the air temperature is high. The following responses, controlled by the hypothalamus, lower body temperature:

(a) **Increased sweating:** sweat glands form sweat from the water, salts and urea they receive from blood capillaries. As the water in sweat evaporates from the skin's surface, so the skin is cooled.

(b) **Vasodilation** (widening of capillaries in the dermis, see Fig. 8.7): the

blood flow to the skin is increased and heat loss by radiation, convection and conduction is increased.
(c) **Reduced muscular activity** (NOTE. Respiration supplies muscles with the energy required for contraction but also produces heat as a by-product).
(d) **Hair lies flat against the skin:** not very important in humans (see below).
(e) **Voluntary actions** may be employed such as removing clothes, reducing artificial heating or switching on a fan.
(f) **Reduced rate of metabolism:** in the long term the appetite may be reduced and less food broken down in respiration or stored.

Failure of the above mechanisms to lower body temperature may result in heat stroke, **hyperthermia**, and then heat death.

2. **Control of overcooling** (blood temperature drops below 37°C):
(a) **Sweating stops.**
(b) **Vasoconstriction** (narrowing of blood vessels in the dermis, (see Fig. 8.7). The blood flow to the skin is decreased and heat loss by radiation, convection and conduction is reduced.
(c) **Increased voluntary muscular activity** increases heat production.
(d) **Increased involuntary muscular activity** (shivering) increases heat production.
(e) **Hair stands on end** ('goose flesh'): this is of little importance in humans, but in other animals with a thick covering of hair (fur) the air trapped between erect hairs significantly reduces heat loss by radiation, convection and conduction.
(f) **Voluntary actions** such as putting on additional clothes or increasing artificial heating may be employed.
(g) **Increased rate of metabolism:** in the long term appetite may be increased and more food broken down by respiration in the liver or stored as fat for insulation.

Failure of the above mechanisms to raise body temperature may result in reduced 'deep body' or 'core' temperature, **hypothermia**, and then cold death.

Definitions

Homeostasis
Excretion
Osmoregulation

Key Words

Liver
Lungs
Kidneys
Skin
Stomata
Excretory products
Nitrogenous wastes
Bile pigments
Ammonia
Urea
Uric acid
Contractile vacuole
Malpighian tubules
Cortex
Medulla
Ureter
Renal artery
Renal vein
Nephron
Pyramids
Pelvis
Glomerulus
Bowman's capsule
Glomerular filtrate
Kidney tubule
First convoluted tubule
Loop of Henlé
Second convoluted tubule
Collecting duct
Urine
Pressure (ultra) filtration
Selective reabsorption
Bladder
Urethra
Diabetes mellitus
ADH (anti-diuretic hormone)
Epidermis
Dermis
Keratin
Germinative layer
Hair follicles
Sweat glands
Sebaceous glands
Sebum
Erector pili muscles
Sense organs
Capillary loops
Radiation
Conduction
Convection
Evaporation
Hypothalamus
Sweating
Latent heat
Vasodilation
Vasoconstriction
Hyperthermia
Hypothermia
Endothermic
Ectothermic
Dialysis

Questions

1. (a) What is excretion?
(b) Name one waste product excreted by both mammals and flowering plants.
(c) Name two other waste products excreted by mammals.
(d) How can leaf fall in some deciduous trees be considered a form of excretion?

2. The diagram below shows a mammalian kidney tubule from Bowman's capsule to one of the collecting ducts.
(a) Describe three *differences between the contents of the tubule at* A *and at* B.
(b) In which part of the kidney would you expect to find
(i) Bowman's capsules?
(ii) collecting ducts?

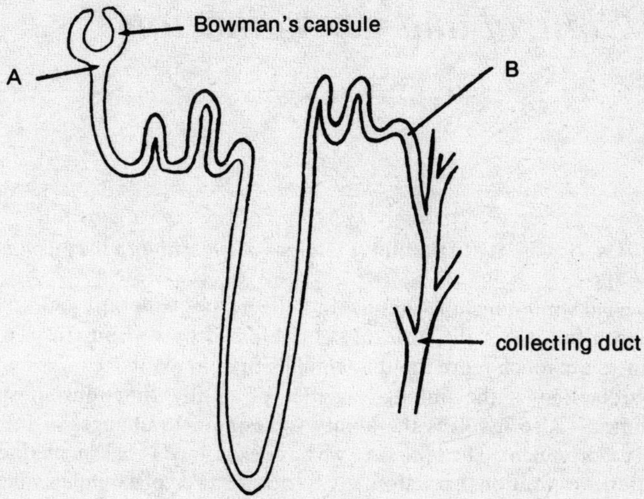

kidney tubule

3. (*a*) *Make a large labelled diagram of a section through mammalian skin.*
(*b*) *Explain how the skin helps mammals to maintain a constant body temperature.*

4. (*a*) *Make a labelled diagram of a section through the mammalian skin.*
(*b*) *For a* named *species of mammal briefly describe how:*
(*i*) *water is gained and lost*
(*ii*) *heat is gained and lost*
(*c*) *How are these changes regulated by the mammal?*

9. *Control and Coordination*

Definition: Sensitivity (irritability) *is an organism's ability to respond to a stimulus.*

External stimuli include changes in temperature, light and sound. The response often takes the form of some kind of movement. In general, responses are much more rapid in animals than in plants.

Coordination is the linking together of bodily functions in space and time. It also involves the ability to respond to changes in the external environment. The processes we have examined in earlier chapters – nutrition, respiration, excretion, etc. – do not take place independently of one another, but are interrelated. For example, aerobic respiration in a contracting muscle would soon cease if the lungs failed to supply oxygen via the circulatory system. During exercise, when the muscles need to gain more oxygen and glucose, and remove excess carbon dioxide, breathing rate is automatically increased and the heart beats faster. This sends a greater volume of oxygenated blood to the muscles. Without coordination, bodily activities would soon malfunction; for example, a runner would collapse after a few metres through lack of an increased oxygen supply to his/her muscles.

The activities of organs and organ systems are not only closely linked to each other and the overall pattern of activity within the body, but to changes outside the body. In response to environmental changes the mammal makes adjustments that tend to keep its internal conditions constant, i.e. homeostasis (page 147). Thus, patterns of activity in the body are responsive to external as well as internal changes.

In mammals and other multicellular animals, coordination is brought about by the action of the **nervous system** (involving electrical **nerve impulses**) and the **endocrine system** (involving chemical messengers called **hormones**). In plants, coordination is under the control of hormones.

9.1. The Mammalian Nervous System

The mammalian nervous system is divided into two parts, the **central**

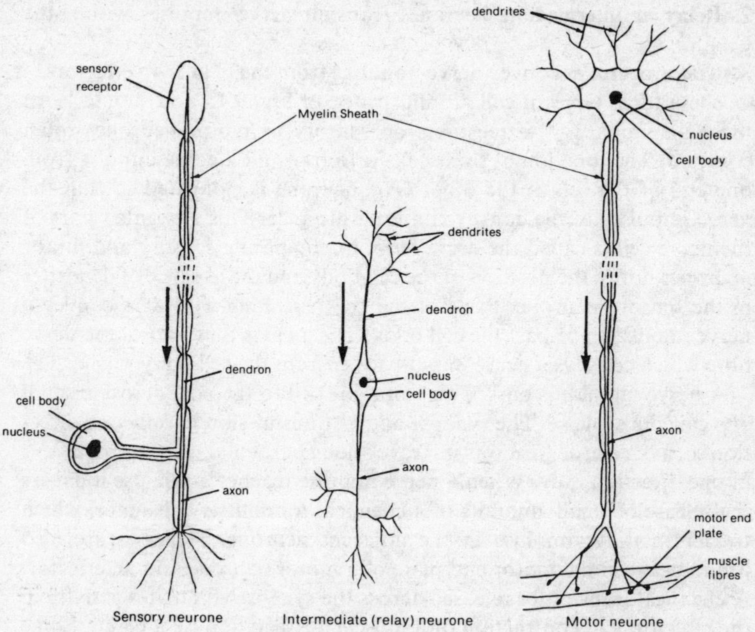

Fig. 9.1. *Three types of nerve cell (neurone)*

nervous system (C.N.S.) – the brain and spinal cord – and the **peripheral nervous system** – the nerves which run into and out of the C.N.S.

Information is transmitted through the nervous system in the form of nerve impulses – essentially weak electric currents. Stimuli from the internal or external environment stimulate **sensory receptors** which convey nerve impulses along **sensory nerve cells** into the C.N.S. The C.N.S. 'processes' this information and may then deliver nerve impulses along **motor nerve cells** to **effectors**, which give an appropriate response. Examples of sensory receptors are sense organs such as the eye (page 172) and the ear and various sensory cells such as pressure receptors in the skin (page 171) and stretch receptors in the muscles (page 172). Muscles and glands are effectors – structures which give a response.

The functional units of the nervous system are nerve cells (neurones). There are three types:

1. **Sensory neurones** convey nerve impulses from sensory receptors to the C.N.S.

2. **Relay or intermediate neurones** transmit nerve impulses within the C.N.S.
3. **Motor neurones** convey nerve impulses from the C.N.S. to effectors.

These three types of cell are illustrated in Fig. 9.1. You should learn these diagrams well – questions on sensory or motor neurones often come up. The function of these cells is to transmit a nerve impulse from one end of the cell to the other. The neurone is elongated so that the nerve impulse can be conveyed a long distance. This elongated part of the nerve cell is called the **nerve fibre**. In comparing sensory and motor neurones notice the position of the cell body and the absence of dendrites in the sensory neurone. Parts of the cell beginning with **den-** convey a nerve impulse to or past the cell body. The **axon** is that part of the nerve fibre which conveys a nerve impulse away from the cell body.

A nerve impulse is passed on from one cell to the next across a small gap called a **synapse**. The synapse allows transmission in only one direction so, as a rule, nerve impulses travel along particular nervous pathways in one direction only. When a nerve impulse reaches a synapse it causes the release of small amounts of a chemical **transmitter** substance, which triggers a nerve impulse in the adjacent neurone. Synapses are also found between the motor end plate of a motor neurone and an effector. A chemical transmitter released across the synapse stimulates activity in the effector, e.g. contraction of a muscle, or secretion by a gland. Some transmitter substances also act as hormones, e.g. adrenalin (page 176).

Activities within the mammalian nervous system are incredibly complex and only the very simplest stimulus–response reactions can easily be described. Reflex actions are such examples.

9.2. *Reflex Action*

Simple reflexes

Definition: A reflex action *is a rapid automatic (involuntary) response to a stimulus.*

Examples of reflexes in man are coughing, sneezing, blinking, sweating, the knee jerk reflex and removing a hand from a hot object. Reflex actions are involved in **homeostasis**, i.e. keeping conditions in the body constant, and in preventing damage to the body. For example, changes in breathing rate during exercise occur automatically without conscious control. This reflex action is a homeostatic function: it serves to keep the blood carbon dioxide level approximately constant. The rapid removal of a hand from a hot object occurs without conscious control. If you waited to 'think' about what response to give, your hand could well be

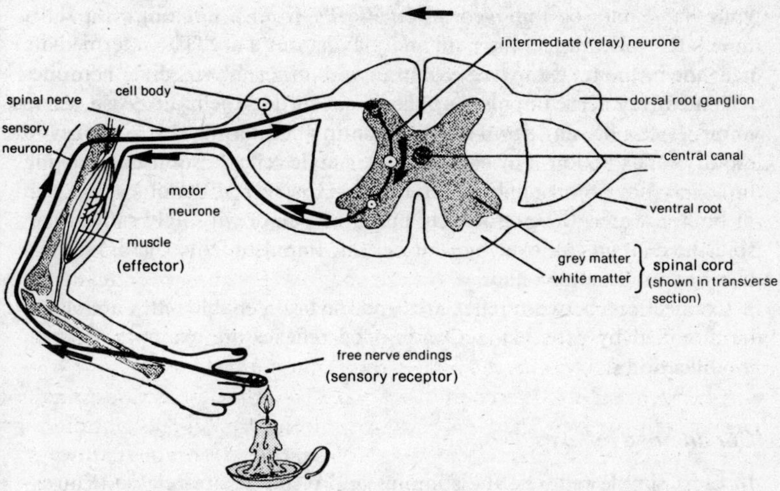

Fig. 9.2. *A reflex arc for the hand withdrawal reflex. → indicates direction of travel of nerve impulse.*
NOTE. *Structures not drawn to scale*

burnt. Instead, the brain is bypassed and preventative action taken before the brain 'decides'.

In reflex action, the nervous pathway between receptor and effector is known as a reflex arc. Simple reflex arcs normally have five components: a sensory receptor, a sensory neurone, an intermediate neurone, a motor neurone, and an effector. The reflex arc for the hand withdrawal reflex is shown in Fig. 9.2.

Study this diagram carefully: in an exam you might have to draw this, or a similar reflex arc, or you may have to label or comment on one provided. Note the positions of neurones and their cell bodies. The sensory neurone runs through the dorsal root of the spinal cord whereas the motor neurone runs through the ventral root. The grey matter of the spinal cord contains a high concentration of cell bodies (it is the nuclei which make grey matter 'grey'), while the white matter contains nerve fibres. You should be aware that the diagram is a simplification. Hundreds of reflex areas are involved in the reflex withdrawal of a hand from a hot object.

Reflexes which involve the spinal cord only and not the brain, are called **spinal reflexes**. Although the brain is not directly involved in these reflex arcs, it does not mean the brain is not informed of events. Although your hand may be removed automatically from a hot object, a short time later you will feel the pain and may let out a cry. The intermediate neurone in the reflex arc has synapses with other intermediate neurones which convey nerve impulses up the spinal cord to the brain. Also, nerve impulses can be sent down from the brain via intermediate neurones to modify reflex action. For example, the simple reflex response to picking up a very hot object might be to drop it. However, if the object is a pan of boiling water, the person may place the pan down safely rather than spill the contents all over him- or herself. Impulses from the brain have overriden the simple reflex.

Connections between reflex arcs and the brain enable reflex actions to be modified by experience. Conditioned reflexes are examples of such modifications.

Conditioned reflexes

In most simple reflexes, the stimulus and response are related. For example, the chemical stimulus of food in the mouth triggers the salivary reflex (secretion of saliva). After a period of learning or training, however, it is possible for a different and often irrelevant stimulus to produce the same response. In such a case a 'conditioned reflex' has been set up and the animal is said to be conditioned to the new stimulus. The classic experiments on conditioned reflexes were carried out by the Russian biologist Pavlov at the beginning of this century.

In dogs, the salivary reflex is triggered by the smell or taste of food. For several days, Pavlov rang a bell at the time food was given to dogs. Later, the sound of the bell alone was a sufficient stimulus to cause a dog's mouth to water, without sight or smell of the food. The original chemical stimulus of the food had been replaced by an unrelated stimulus detected by the ears.

The training of animals is done largely by conditioning them to respond to new stimuli. To what extent conditioned reflexes play a role in human behaviour is difficult to say. However, the learning of habits and basic skills probably occurs in a similar fashion. For example, walking, running, cycling, using a knife and fork, etc., although all originally learned consciously, later function below the level of consciousness. However, these patterns of behaviour are clearly much more complex than simple conditioned reflexes.

The advantage of simple and conditioned reflex responses is that they leave the conscious mind free to think about other less routine matters which do require concentration and reasoning.

9.3. *The Central Nervous System*

The spinal cord

The spinal cord is a hollow tube comprising white matter (nerve fibres) on the outside, grey matter (cell bodies) in an H-shaped zone on the inside, and a central canal containing nutritive fluid (see Fig. 9.2); memorize this arrangement.

The spinal cord runs through, and is protected by, the vertebral column (see page 193). At regular intervals paired, spinal nerves emerge from the spinal cord. These nerves typically contain both sensory and motor nerve fibres (see Fig. 9.2) enclosed in a connective tissue sheath.

NOTE. A **nerve** contains many hundreds of **nerve fibres** – do not confuse the two.

The spinal cord has three main functions:

1. It contains the circuitry for reflex arcs involving the muscles or glands of the body, as distinct from those of the head.
2. It provides connections between reflex arcs, both along and across the cord, so that a limited degree of coordination is possible.
3. It acts as a relay station for impulses passing between the brain and the lower parts of the body.

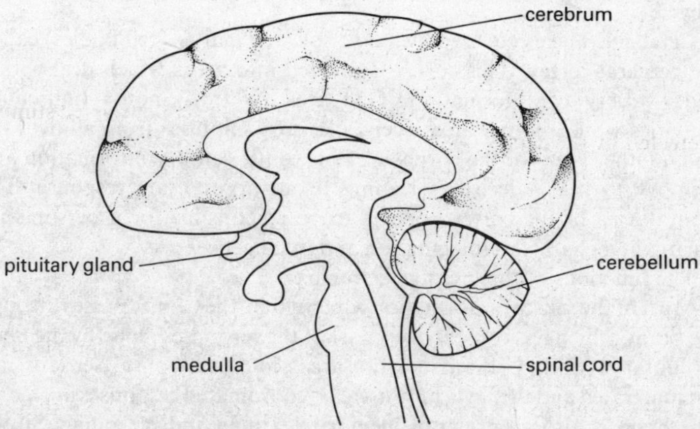

Fig. 9.3. *The human brain*

Brain

The brain is the enlarged front end of the spinal cord in vertebrates. It acts as a centre for reflex arcs in the head involving the main sense organs, the eyes, ears, mouth and nose. The increased importance of these sense organs has led, during evolution, to an increase in the size and complexity of the brain.

Like the spinal cord, the brain consists of nerve cells with a great concentration of cell bodies. In simpler vertebrates such as fish and amphibians, three regions of the brain are distinguishable: the **fore-**, **mid-** and **hind-brain**. The fore-brain receives nerve impulses from the nasal organs, the mid-brain receives impulses from the eyes, while the hind-brain receives impulses from the ears and sensory receptors in the skin. In mammals, these three regions are not clearly distinguishable due to the great expansion of the fore-brain to form the **cerebral hemispheres** or **cerebrum**. In man, this is the largest part of the brain and spreads out and covers most of the brain.

The major parts of the human brain are as follows (see Fig. 9.3).

Medulla. This part of the brain coordinates the unconscious (involuntary) activities of the body (e.g. rates of heartbeat and breathing, control of body temperature and blood pressure).

Cerebellum. The cerebellum deals with reflexes. It receives messages from the organs of balance in the inner ear and from stretch receptors in the joints and muscles. This information is used to achieve balance and muscular coordination in activities like walking, running and riding a bicycle.

Cerebum. This is covered by a layer of grey matter (cell bodies) called the **cerebral cortex**. This cortex is divided into areas which deal with a wide variety of information. One area controls motor (muscular) activities while another area receives sensory impulses from all over the body. Other areas of the cortex are able to use sensory information and memory to make decisions leading to an appropriate response. The frontal area of the cortex appears to be responsible for a person's individuality, character, imagination and intelligence.

The functions of the brain, in summary:

Most of the sensory information which enters the C.N.S. passes through the brain, which has control over most of the responses made by the body (voluntary action). Using association areas, impulses from all sense organs are integrated and then, via motor areas, coordinated responses are made. The brain is also an organ of memory, learning and reasoning. Stored information can be used to modify behaviour according to past experience.

9.4. *Sensory Receptors: Sensory Cells and Sense Organs*

Sensory receptors are specialized cells which convert stimuli into nerve impulses. These cells are the point of entry for information entering the nervous system. Sensory receptors may consist of isolated single cells, e.g. pressure receptors in the skin (page 157), or they may be collected together to form a highly efficient sense organ, e.g. the eye (page 172).

In mammals, it is only in the brain that a stimulus is interpreted (perceived) and identified. For example, if the nerve running from the eye is damaged, the person may be blind although the eye itself may function perfectly.

1. **External receptors** detect stimuli arising from outside the body. The main ones are:

(a) **Skin receptors**, which detect heat, cold, touch, pressure and pain (see Fig. 8.6, page 157). Certain regions of skin are more sensitive to particular stimuli than others. For example, the tip of the tongue, the lips and fingertips are very sensitive to touch. In these areas the appropriate sensory cells each have their own nerve fibre running to the C.N.S. In other less sensitive areas, several sensory cells share the same nerve fibre.

(b) In the nasal cavity, **olfactory organs** are sensitive to smell. They detect chemical substances which dissolve in the film of moisture covering the sensory cells.

(c) On the tongue, **taste receptors** are grouped together in **taste buds** which, like olfactory organs, detect chemical substances which dissolve in the film of moisture covering them. There are four types of taste buds sensitive to bitterness, saltiness, sourness and sweetness.

(d) In the retina of the eye, **light receptors**, rods and cones, are sensitive to different wavelengths and intensities of light (see page 175).

(e) In the inner ear, the **organ of Corti** in the cochlea is sensitive to vibrations (sound). **Sensory hair cells** in the sacculus, utriculus and semicircular canals are sensitive to the orientation and movement of the body.

2. **Internal receptors** respond to stimuli within the body and are often associated with homeostatic mechanisms (see pages 147–8);

(a) the **hypothalamus** detects changes in blood temperature.

(b) the **hypothalamus** detects changes in blood osmotic potential (see page 42).

(c) The **respiratory centre** of the medulla oblongata is sensitive to carbon dioxide level.

(d) **Proprioreceptors** respond to stimuli from muscles and joints. One type, stretch receptors (page 200), are embedded in every skeletal muscle in the body, and similar receptors are found in muscle tendons, and in the ligaments of joints (page 191). These receptors provide the brain with information about the degree of tension in each part of the muscular system, and the angle to which every joint is bent. Without proprioreceptors, muscular coordination and locomotion would be impossible.

The human eye

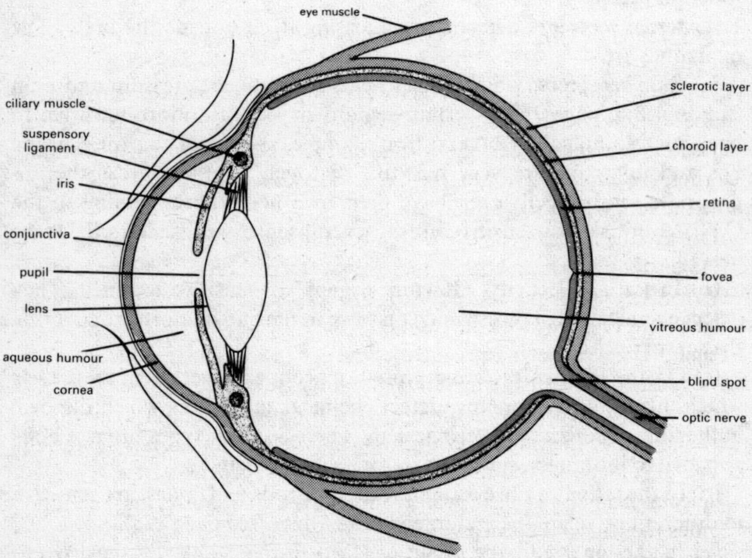

Fig. 9.4. *V.S. of the human eye*

The human eye is a remarkable device for focusing light rays and converting them into nerve impulses. These impulses are carried along nerve fibres to the brain where an image of remarkable precision is produced.

You are unlikely to be asked to draw a detailed fully labelled diagram of the human eye (see Fig. 9.4). However, you may well be given a diagram which you must label. Alternatively, you may have to draw simplified diagrams to illustrate how the eye works.

The eye functions by focusing an image on the light-sensitive layer, the retina, at the back of the eye. Light entering the eye takes the follow-

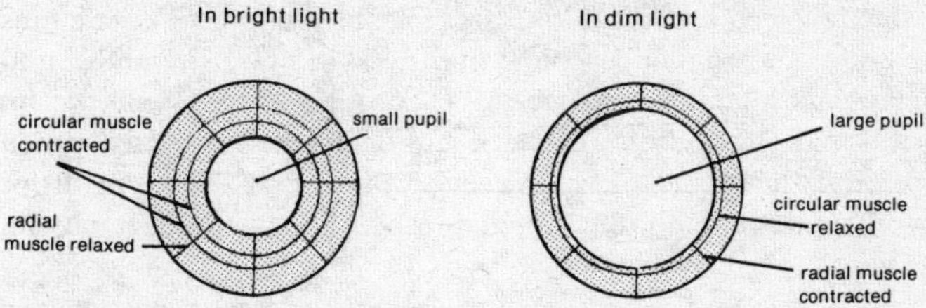

Fig. 9.5. *Altering the amount of light entering the eye: the iris/pupil response to light*

ing path: conjunctiva → cornea → aqueous humour → pupil → lens → vitreous humour → retina.

To help you remember the structure of the eye, you can think of it as a number of layers. The conjunctiva (the outer covering of the front of the eye) is joined to the eyelids; the cornea and sclerotic layer form a continuous ring; the choroid layer, iris, ciliary body and lens form another ring, while the retina forms the inner lining.

Here are the parts of the eye with a brief mention of their function:

Conjunctiva: a thin layer of transparent cells protecting the cornea.

Cornea: the transparent part of the sclerotic layer which refracts (bends) light rays towards the retina.

Pupil: the dark spot in the centre of the eye as seen from the front. It is a hole in the iris through which all light enters the eye.

Iris: the coloured part of the eye. It is a muscular diaphragm which alters the size of the pupil and thus adjusts the amount of light entering the eye. It is made up of radial and circular muscles (Fig. 9.5).

Aqueous humour: a clear watery fluid which supports the lens and provides the lens and cornea with food and oxygen.

Lens: made up of transparent living cells. It is elastic and focuses light rays on to the retina to form an image.

Ciliary muscles: circular muscles which alter the shape of the lens during focusing (accommodation).

Suspensory ligaments: hold the lens in place and are attached to the ciliary muscles.

Vitreous humour: clear jelly-like substance which supports the eyeball.

Retina: contains the light-sensitive sensory cells, the rods and cones.

Fovea: the most sensitive part of the retina where most light rays are focused. It contains colour-sensitive cones only.

Blind spot: the point where the optic nerve leaves the eye and blood vessels enter and leave. There are no light-sensitive cells here.

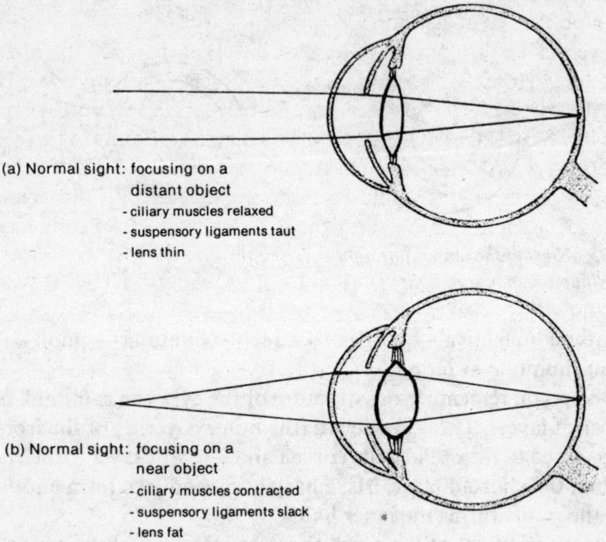

Fig. 9.6. *The human eye: focusing (accommodation)*

Choroid layer: a dark layer which stops light being reflected. It contains blood vessels which supply food and oxygen to other parts of the eye.

Sclerotic layer: the tough outer covering of the eye. It maintains the shape of the eyeball.

Eye muscles: these are attached between the sclerotic layer and the skull. They move the eyeball.

Optic nerve: this contains the bundle of nerve fibres which carry nerve impulses to the brain.

Controlling the amount of light entering the eye

The pupil size is adjusted by reflex action according to the prevailing light conditions (light intensity). In bright light damage to the retina is prevented, and in dim light, maximum use is made of available light.

In bright light, the circular muscle of the iris diaphragm contracts and the pupil becomes smaller (see Fig. 9.5). This prevents too much light entering the eye. In dim light, the radial muscles contract, and the pupil becomes larger, allowing more light to enter (see Fig. 9.5).

Pupil size is also affected by focusing distance (see above).

Focusing (accommodation)

The eye is able to focus on objects at different distances by altering the shape (curvature) of the lens. When an object is a long distance away, the light rays reflected from it are almost parallel when they enter the eye. They have to be refracted (bent) by a small amount to be focused on the retina. A thin lens shape is required. The ciliary muscles relax and the pressure of the eye fluids (humours) pushes out and tightens the suspensory ligaments, and the lens is pulled out into a thin shape (see Fig. 9.6a).

When an object is close to the eye, the light rays reflected from it are diverging (moving apart) when they enter the eye. They need to be refracted to a larger degree to be focused on the retina. A fatter lens shape is required. The ciliary muscles contract, the suspensory ligaments slacken, and the elastic lens springs back to a fatter shape (see Fig. 9.6b).

For technical reasons, the pupil size tends to be smaller when focusing on a close object.

The retina

The retina is made up of two types of sensory cell, **rods** which are sensitive to low light intensities, and **cones**, which respond only to bright light and detect colour. There are thought to be three types of cone cell in mammals, each responding to one of the primary colours, red, green and blue. Light of other colours and shades stimulates varying proportions of these cones. Impulses from these pass to the brain where the sensation of colour is produced. In man, cones are concentrated on the fovea, the most sensitive part of the retina. Nocturnal animals such as owls have many rods and few cones giving them better night vision but poor colour discrimination.

Stereoscopic (binocular) vision

Stereoscopic or three-dimensional vision occurs when both eyes focus on the same object. Each eye receives a slightly different view of the object and from these the brain forms a three-dimensional impression. Stereoscopic vision enables distance and depth to be judged accurately. Humans and predatory animals such as cats and owls have well-developed stereoscopic vision. Their eyes are placed on the front of the head looking forward. Prey animals such as the rabbit have their eyes on the sides of the head so that they have a wide field of vision for detecting an approaching predator. They have little or no stereoscopic vision.

9.5. *The Mammalian Endocrine System*

Table 9.1 *Differences between nervous and hormonal action*

	Nervous	*Hormonal*
1. Method of transmission	Electrical impulse passing along nerve fibre (chemical transmission across synapses)	Chemical passing through bloodstream
2. Speed of transmission	Rapid	Slow
3. Speed of response	Immediate	Usually slow, e.g. thyroxine, but may be fast, e.g. adrenalin
4. Duration of response	Short-lived	Longer-lasting
5. Localization of response	Response very localized	Response sometimes widespread, e.g. adrenalin

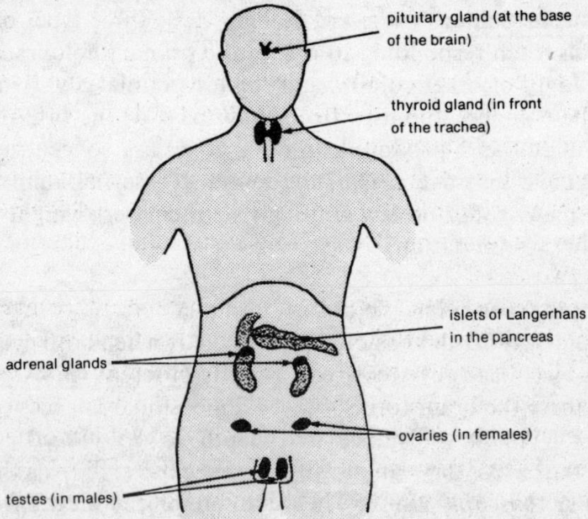

Fig. 9.7 *The main endocrine glands in the human body*

The **endocrine system** is made up of **endocrine (ductless) glands** which secrete their products, **hormones**, in minute quantities directly into the

Table 9.2. *Summary of some of the endocrine glands and hormones in humans*

Endocrine gland	*Hormone*	*Stimulus*	*Effect*
Thyroid	thyroxine	Hormone from pituitary	Increases metabolic rate.
Islets of Langerhans in the pancreas	insulin	High blood glucose level	Lowers blood glucose levels; stimulates glucose uptake by cells.
	glucagon	Low blood glucose level	Raises blood glucose levels; stimulates release of glucose from cells.
Adrenal glands	adrenalin	Danger, fright, stress	Increases heartbeat and breathing rate; diverts blood from gut to brain and muscles; raises blood glucose levels.

(NOTE. Check your syllabus to see which of these you need to know about.)

blood, rather than through tubes (ducts). These hormones travel through the bloodstream and exert an effect on **target** organs, tissues or cells, by speeding up or slowing down their activity.

The endocrine system, sometimes working in conjunction with the nervous system, controls and coordinates growth, development and activity within the organism. The nervous system and endocrine system are similar in two ways:

1. They are both set into action by a stimulus to produce a response.
2. They both involve chemical transmission. In the case of the nervous system this occurs only across synapses, whereas in the endocrine system the chemical, a hormone, travels considerably further.

More usually, the *differences* between nervous and hormonal action are stressed. These are summarized in Table 9.1.

The main endocrine glands in the human body are the pituitary, thyroid, adrenals, pancreas and the reproductive organs. The positions of these glands are shown in Fig. 9.7. The pituitary gland, sometimes called the 'master gland', coordinates and controls the activity of the other glands. It releases hormones which stimulate these glands to release their own hormones.

The various endocrine glands are listed in Table 9.2 together with their hormones, the stimulus which triggers release of the hormone, and the effect of the hormone on target tissue. Knowledge of these glands, their hormones and their effects is frequently demanded in multiple-choice or short-answer papers.

The level of circulating hormone is controlled by a **'feedback'** process; a hormone adjusts its own output by affecting the endocrine gland that causes its secretion.

The brain monitors the level of hormone in the blood. If too much is present the pituitary gland is blocked from releasing the appropriate stimulating hormone; if too little is present, more stimulating hormone is released. Feedback control provides an effective means of monitoring and maintaining suitable hormone levels in the body.

In the human female menstrual cycle (page 232), feedback of several hormones generates cyclical changes. Several hormones often affect the level of a substance or an activity in the body. For example, blood glucose level is affected by three hormones: insulin, thyroxine and adrenalin.

9.6. *Homeostatic Mechanisms*

In Chapter 8 we looked at the concept of homeostasis – keeping conditions within the body constant. The mechanisms involved in homeostasis all follow the same basic plan and involve either nervous or hormonal action, or a combination of the two. This basic plan can be represented as follows:

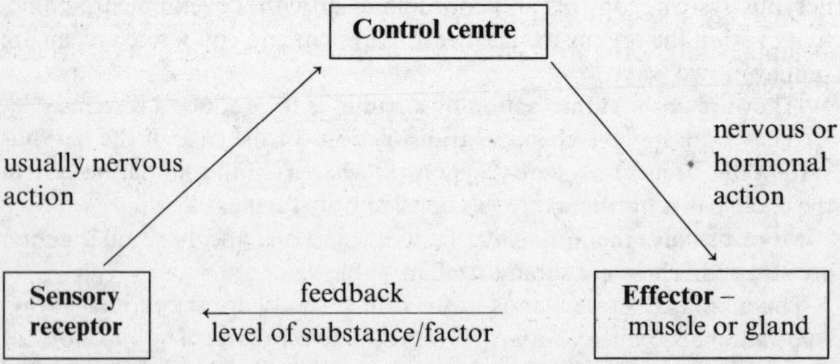

As with the control of hormone levels in the blood, the level of other substances and factors in homeostasis is controlled by feedback.

To understand the above scheme it is best to apply it to a few examples.

Once understood, this scheme is very useful in checking your knowledge and understanding of different homeostatic mechanisms. It is also a convenient way of remembering them.

1. **Control of breathing:** keeping the level of carbon dioxide in the blood constant. If the level of carbon dioxide in the blood rises this is sensed by the respiratory centre in the medulla oblongata of the brain (page 170). This sends out nerve impulses to the diaphragm and intercostal muscles which contract more frequently and more strongly, increasing the rate and depth of breathing (see page 108). As a result, excess carbon dioxide in the blood is more rapidly removed from the body. When the level of carbon dioxide returns to normal, this is sensed by the respiratory centre. It sends out fewer nerve impulses to the diaphragm and intercostal muscles, and as a result the rate and depth of breathing return to normal.

The respiratory centre is the sensory receptor and control centre, while the diaphragm and intercostal muscles are the effectors:

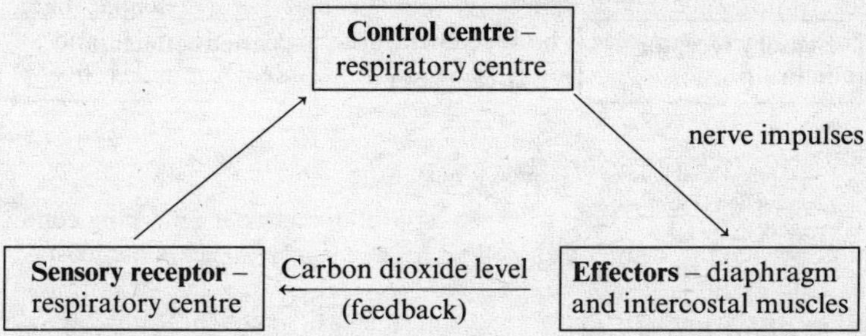

2. **Control of blood glucose level** (see page 78): where blood glucose levels are high, the hormone insulin is released by the pancreas. This stimulates the uptake of glucose by cells in the body, and the conversion of glucose to glycogen by the liver and muscles.

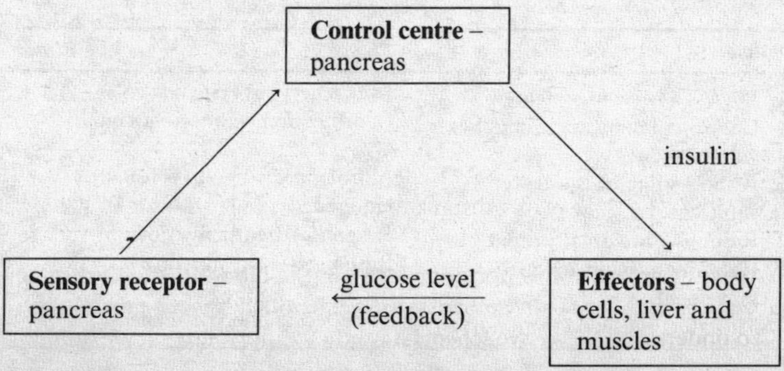

3. **Control of body temperature** (see page 159): when body temperature rises above normal this is sensed by the brain which sends nerve impulses to various effectors in the skin which give responses to increase heat loss and thus lower body temperature (see page 159). When the body temperature drops below normal, the brain sends nerve impulses to effectors in the skin which respond by reducing heat loss. At the same time skeletal muscles may be stimulated to contract (shivering).

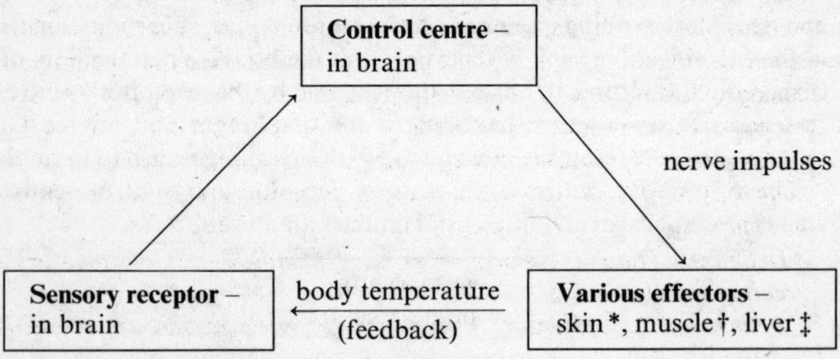

* sweating, vasodilation and vasoconstriction
† contraction (shivering)
‡ metabolic rate

9.7. *Plant Responses*

Plants are able to respond to light, gravity, temperature, chemical substances, water and, in some cases (such as climbing plants), to touch.

Table 9.3. *Differences between animal and plant responses*

Animals	*Plants*
1. Possess a nervous system	Lack a nervous system
2. Coordination by nerve impulses and hormones	Coordination by hormones only
3. A short stimulus is usually sufficient	A prolonged stimulus is usually required
4. Responses are usually rapid	Responses are usually slow
5. Responses usually involve movement	Responses usually involve growth
6. The effect is usually temporary	The effect is often permanent

Plant responses are due to different rates of growth, as in the bending of a shoot towards light or, less often, to changes in turgidity, as in stomatal pores or the petals of a flower opening and closing.

Most plant responses can be classed as either **nastic responses** or **tropic responses.**

Definition: Nastic responses *are movements by part of the plant in which the direction of response is not determined by the direction of the stimulus.*

Examples of nastic responses are the opening of the petals of the daisy flower during daylight and their closure at night, and the opening and closing of stomata with changes in light intensity (see page 140). Such responses are non-directional.

Definition: Tropic responses (tropisms) *are growth movements by the plant in which the direction of movement is determined by the direction of the stimulus.*

Definition: Phototropism *is a plant growth movement in response to the direction of light.*

Definition: Geotropism *is a plant growth movement in response to the direction of the pull of gravity.*

Tropisms are either **positive**, when movement is towards the source of the stimulus, or **negative**, when movement is away from the source of the stimulus.

Main stems, for example, are **negatively geotropic** (they grow upwards, away from the pull of gravity), while main roots are **positively geotropic** (they grow downwards, towards the pull of gravity). A plant stem growing towards a light source is **positively phototropic**.

Tropisms are survival responses. For example:

1. They enable seedlings to become established with the roots in the soil and leaves in the air.
2. They enable a plant to change its growth pattern at any time in its life to suit changing circumstances. For example, if a plant survives being blown over in the wind, new growth will make the stem curve upwards under the influence of negative geotropism.

Tropisms may be investigated experimentally using seedlings, since these are particularly sensitive to the stimuli of light and gravity. The **coleoptiles** of oat or maize are traditionally used because they do not have leaves or buds to confuse findings.

Phototropism experiments

A simple experiment to determine the effect of light on the growth of oat

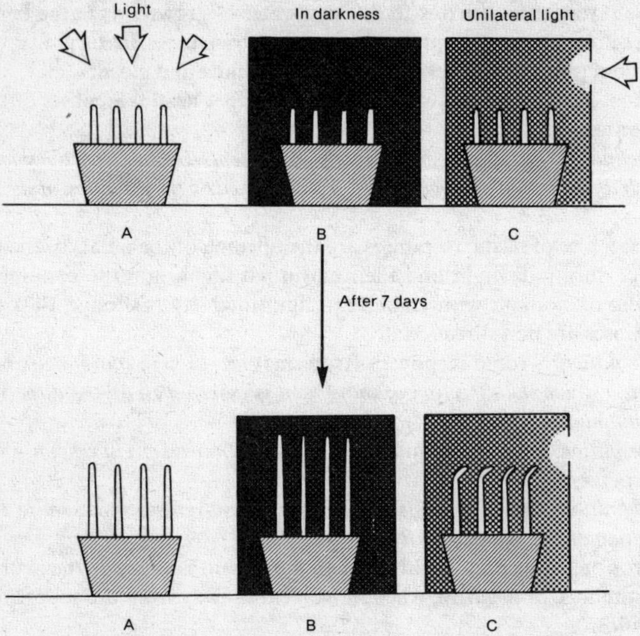

Fig. 9.8. *Experiment to determine the effect of light on oat seedlings*

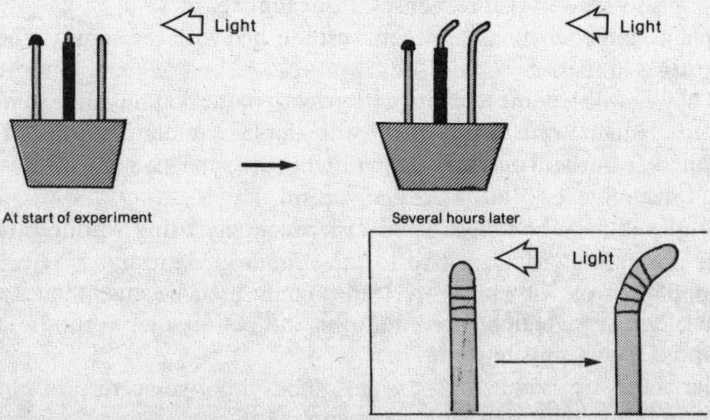

Fig. 9.9. *Investigations of regions of sensitivity and response in a coleoptile*

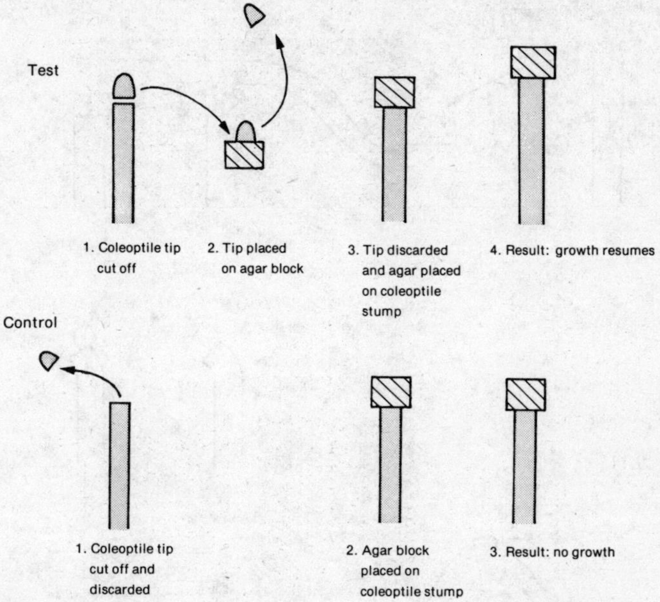

Fig. 9.10.

seedlings is shown in Fig. 9.8. As far as possible, conditions are the same in all three pots, except for the presence, absence or direction of light.

The seedlings in C receive unilateral light (light from one side) and show a growth curvature towards the light. The seedlings in A show normal growth. The seedlings in B, without light, grow tall and spindly. They are yellow and lacking in chlorophyll. This type of response to lack of light is termed **etiolation**. It is a good example of a survival response. In the wild, if the seed lands in dense undergrowth where it is shaded from light, it will grow rapidly upwards to reach the light, where it can start to photosynthesize and manufacture its own food. Without light, the seedling cannot photosynthesize and will eventually die.

The experiment shown in Fig. 9.9 can be used to investigate the region of sensitivity and region of response in the phototropic response of a coleoptile. The results indicate that:

1. The shoot tip is sensitive to light. The phototropic response only occurs in those shoot tips which are exposed to light.
2. The region of growth response (bending) occurs immediately behind

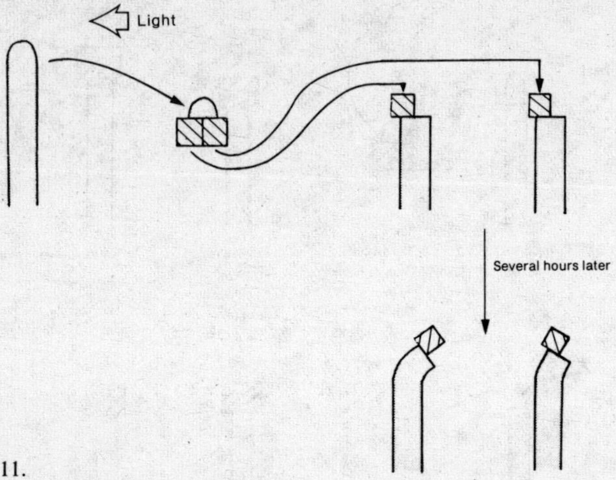

Fig. 9.11.

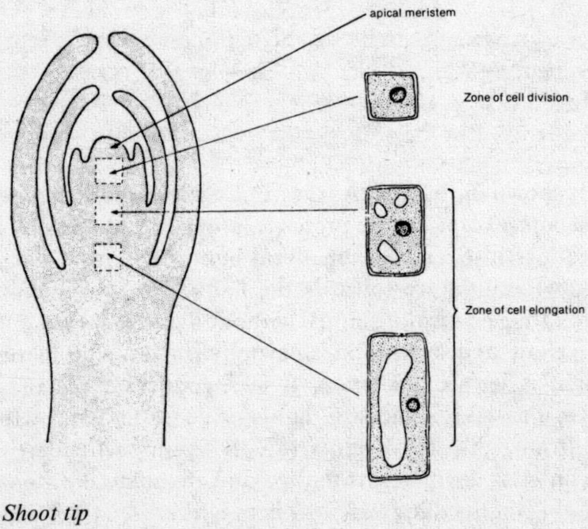

Fig. 9.12. *Shoot tip*

the tip in the region of elongation. This can be better shown by marking this region with lines (see inset, Fig. 9.9).

3. There must be some 'influence' passing down from the tip to the region of cell elongation. The experiment does *not* indicate what this influence is (be careful not to draw conclusions not warranted by the results).

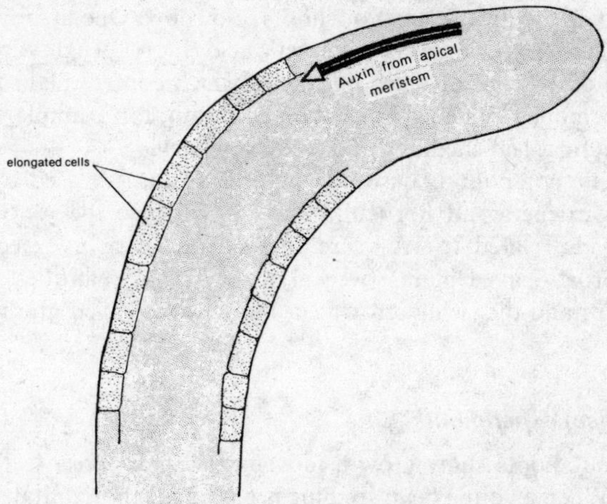

Fig. 9.13. *Cell elongation causing a growth curvative towards the light*

Further experiments can be conducted to investigate this influence. In the experiment in Fig. 9.10 the tip of a coleoptile is removed and placed for several hours on an agar block. The block is then placed on the coleoptile stump. For the control, the process is repeated but the coleoptile tip discarded and an untreated agar block placed on the stump. The results indicate that some substance present in the coleoptile tip, and transferred to the agar block, stimulates growth. A normal agar block used as a control does not promote growth.

The above experiment indicates the presence of a growth-stimulating substance in the coleoptile tip. The next experiment (Fig. 9.11) indicates that such a substance is responsible for growth curvatures in response to light. The agar block from the side of the shoot furthest from the light causes a greater curvature than the block from the side nearest the light. The generally accepted conclusion is that light somehow causes more of a growth-stimulating chemical to gather on the side of the coleoptile tip furthest from the light, i.e. on the shaded side.

There is now strong evidence that in many plants a chemical called **auxin** is produced by the region of actively dividing cells at the tip of the shoot (the apical meristem) (see Fig. 9.12). More auxin passes down the shaded side of the shoot where it stimulates growth of cells in the region of cell elongation, causing them to elongate. The result is a growth curvature towards the light (see Fig. 9.13).

The general name auxin is given to those growth substances which are produced at the apices (tips) of shoots and roots. One auxin which has been isolated from plants is indoleacetic acid (IAA). Auxins are the most common of a range of plant hormones which control plant responses. Other hormones influence not only growth but, for example, flowering, bud sprouting, leaf shedding and seed dormancy.

Synthetic compounds based on natural auxins are used by farmers, market gardeners and horticulturalists to increase the yield of crops. Compounds related to IAA are sprayed on lawns as selective weed killers; broad-leaved plants (weeds) absorb large quantities, 'outgrow' themselves and die, while growth in the narrow-leaved grasses is stimulated.

Geotropism experiments

Roots and shoots show growth movements in response to the pull of gravity. When a young bean seedling is placed in a horizontal position in the dark, and well supplied with water, growth of the shoot curves upwards and root growth curves downwards (see Fig. 9.14). The shoot is negatively geotropic and the root is positively geotropic.

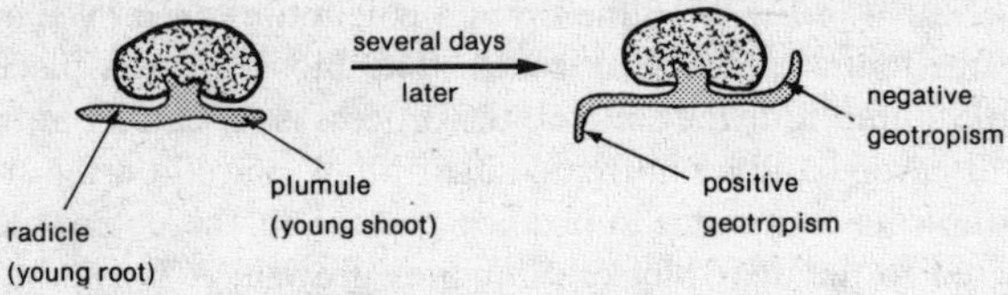

Fig. 9.14. *Geotropism in a germinating bean*

In performing experiments to examine the effect of the pull of gravity, a control is established using a **clinostat** (a slowly rotating drum). This rotates the seedling, subjecting it to the pull of gravity equally on all sides. The results of a typical experiment are shown in Fig. 9.15).

In geotropic responses, auxins are thought to migrate to the lower side of root and shoot under the pull of gravity. In the root, auxins inhibit cell elongation, thus causing the root to bend downwards. In the shoot, auxins stimulate cell elongation, thus causing the shoot to bend upwards.

Artificial uses of plant hormones

Some plant hormones can be made artificially and are used for a wide variety of purposes in agriculture and horticulture:

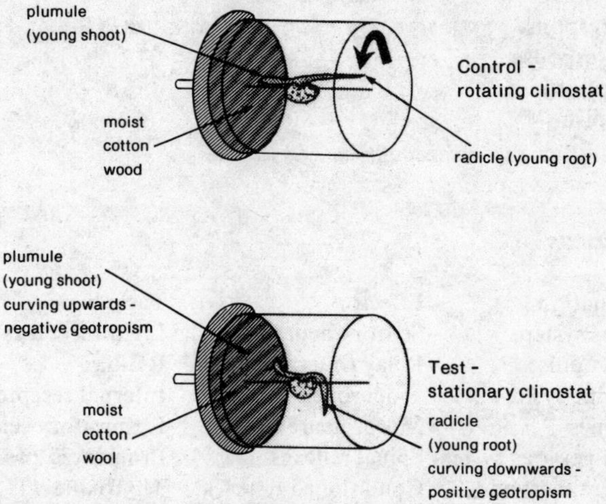

Fig. 9.15. *Investigation of geotropism in a germinating bean*

1. **Weed killers** High concentrations of certain hormones make plants grow so fast that they run out of raw materials and die.
2. **Initiating root development** It is possible to grow a new plant from a cutting from another plant, usually a leaf or part of a stem. A hormone is added to the base of the cutting to encourage it to develop roots.
3. **Stimulating fruit development** Seedless grapes are produced by treating them with a hormone which stimulates development of the fruit in the absence of seeds.
4. **Prolonging shelf life of crops** Potatoes are sprayed with a hormone to inhibit sprouting. This allows them to be kept longer in the shops. The life of fresh leaf crops such as lettuce is also prolonged by the use of artificial hormones. Flowers are treated in a similar way.
5. **Ripening** e.g. of tomatoes and citrus fruits can be controlled by the use of hormones.
6. **Fruit fall** The time at which fruits fall from trees can be controlled with hormones. This allows growers to choose their picking time.

Definitions

Sensitivity (irritability)
Reflex action

Nastic response
Tropic response
Phototropism
Geotropism

Key words

Coordination
Nervous system
Nerve impulses
Endocrine system
Hormones
Central nervous system
Peripheral nervous system
Autonomic nervous system
Cerebellum
Thalamus
Cerebral cortex
Motor areas
Sensory areas
Association areas
Growth hormone
Sensory receptors

Effectors
Sensory neurones
Relay (intermediate) neurones
Motor neurones
Spinal reflexes
Conditioned reflexes
Spinal cord
White matter
Grey matter
Central canal
Brain
Cerebrum
Medulla oblongata
External receptors
Skin receptors
Olfactory organs
Taste receptors

Taste buds
Light receptors
Retina
Internal receptors
Respiratory centre
Proprioceptors
Thyroxine
Insulin
Adrenalin
Feedback
Coleoptile
Etiolation
Agar
Meristem
Cell elongation
Auxin
Indoleacetic acid (IAA)

Questions

1. Sensitivity (irritability) may be defined as the organism's ability to respond to a stimulus. The diagram below summarizes the mechanism of most such responses.

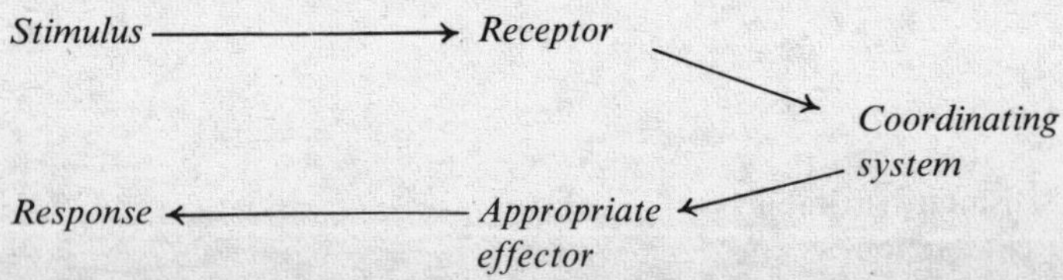

(*a*) *Explain how the following responses fit into this diagram.*
(*i*) *a person dropping a hot plate.*
(*ii*) *a plant growing towards light.*
(*b*) *The flowers of some plants close at night and open during the day. Suggest two stimuli which would cause this response.*

2. (*a*) *Make a labelled diagram to show a transverse section of a mammalian spinal cord.*
(*b*) *Add to your diagram the neurones involved in a reflex arc, showing the directions travelled by the impulses.*

3. (*a*) *Describe an experiment you might do to find out which part of a growing shoot responds to light.*
(*b*) *Describe what control experiments you would do and explain their purpose.*

10. Support and Movement

All organisms have to support themselves, to a greater or lesser degree, in order to maintain body shape. Self-support is usually achieved using some kind of skeleton. The degree to which an organism must support itself is largely determined by the environment it lives in. In aquatic environments, the water provides a high degree of support and less self-support is required. Many aquatic invertebrates do not possess a skeleton.

On land, the air provides little support and much more self-support is required. Land vertebrates depend almost entirely on their skeleton for support; it raises the body off the ground and makes locomotion (movement from one place to another) possible. In addition, many of the internal organs are suspended from the skeleton. Flowering plants on land are supported by the turgor of packing tissues (page 43) and by rigid supporting tissues such as xylem vessels (page 136).

Movement is a characteristic feature of living organisms, but is much more obvious in animals than in plants. Animals, in general, can move more rapidly and most show locomotion. Only a few simple plants, e.g. *Chlamydomonas* (page 297), show locomotion, and most plants move only part of the body, and then only slowly (see page 180).

The difference in locomotory ability between animals and plants is related to their method of nutrition. On land, green plants need simply to bask in the sun and obtain water and minerals from the soil in order to feed (see page 48). Animals, on the other hand, are heterotrophic (page 47) and require ready-made organic food. They may have to travel long distances to find a suitable source. Being freely mobile has other advantages: they can avoid predators, seek out mates, and disperse to new areas.

10.1. *Types of Skeleton*

With only a few exceptions, **locomotion** is brought about by contraction of muscles acting against some kind of skeleton. In some simpler organisms, e.g. *Amoeba* (page 315), locomotion does not involve a skeleton in the normal sense.

All **skeletons** are resistant to compression, and provide a rigid framework for the body. Three types of skeleton are found in animals and they are often important in protection as well as support and locomotion.

1. A **hydrostatic** skeleton is found in the earthworm (page 317). It is a fluid skeleton and as such is incompressible (does not change in volume when a pressure is applied), but it does change shape. Muscles contract against the fluid skeleton and cause changes in body shape which give rise to locomotion. The pressure of fluid pressing out against the muscles supports the body.
2. An **exoskeleton** (outer skeleton) is found in arthropods, e.g. insects (page 319). This rigid outer covering is called a **cuticle**. At the joints the skeleton is soft and flexible to allow for movement. Limbs are jointed and hollow and have inward projections for muscle attachment. To allow for growth, the exoskeleton must be periodically shed (ecdysis) and a new one laid down.
3. An **endoskeleton** (internal skeleton) is found in vertebrates. It is made of hard material, either **bone** or **cartilage**. The limbs are jointed, and muscles are attached to the bones, which act as levers.

10.2. *The Mammalian Skeleton*

The skeleton of mammals is composed largely of bone. Bone is not dead material, but is made up of a non-living part, mostly calcium phosphate, together with the protein collagen, secreted by living bone cells. The cells are found in spaces within the non-living part, and they receive food and oxygen from blood vessels which run through the bone. Apart from its role in support and movement, bone is the site where red and white blood cells and platelets are made (in the bone marrow, see page 117). Bone is also the site of calcium storage. If the body tissues are short of calcium, bone cells release some calcium salts into the blood. They are replaced when the body's intake of calcium is resumed.

Cartilage is found closely associated with bone. It is softer than bone and is found at joints where one bone articulates with (moves against) another. It is a smooth material and produces very little friction at points of contact. Its ability to 'give' slightly when compressed means it has a cushioning effect and absorbs shocks.

Associated with the bones of the skeleton are **ligaments** (which attach bone to bone), **tendons** (which attach muscle to bone) and **muscles**, which when they contract move one bone relative to another.

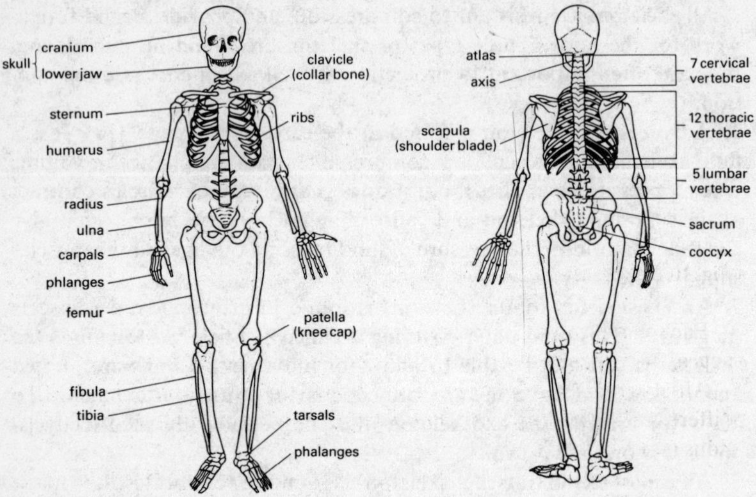

Fig. 10.1. *The human skeleton*

The mammalian skeleton has two parts:

1. The **axial** skeleton consists of the skull, vertebral column (backbone), ribs and sternum (breast bone).

The **skull** consists of the cranium (brain box) and jaws. The cranium is made of a number of fused bones which enclose and protect the brain and organs of hearing, sight and smell. The upper jaw is fixed to the cranium, while the lower jaw is jointed to it.

The **vertebral column** in man is made up of 33 bones called **vertebrae** (singular: vertebra). The vertebral column performs three main functions:

(a) It protects the spinal cord (see page 169).

(b) It enables flexible movement of the body.

(c) It determines the body's posture.

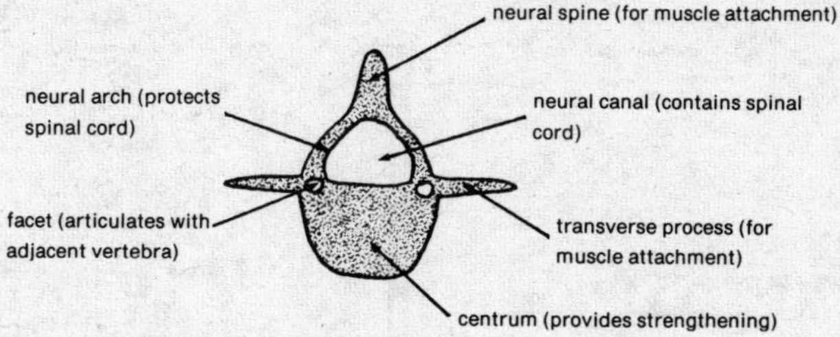

Fig. 10.2. *Generalized structure of a vertebra*

Each vertebra is separated from the next by a disc of cartilage which acts as a shock absorber and allows limited movement.

Although all vertebrae have the same basic structure (Fig. 10.2), in different regions of the backbone they are specialized for different functions (see Fig. 10.3):

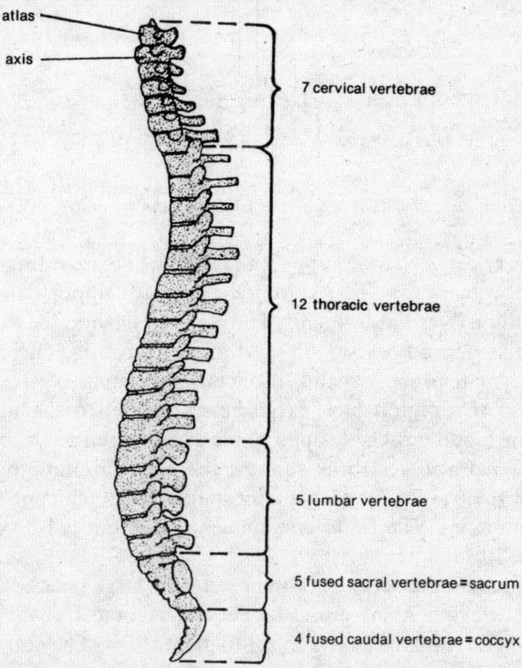

Fig. 10.3. *Side (lateral) view of the human vertebral column*

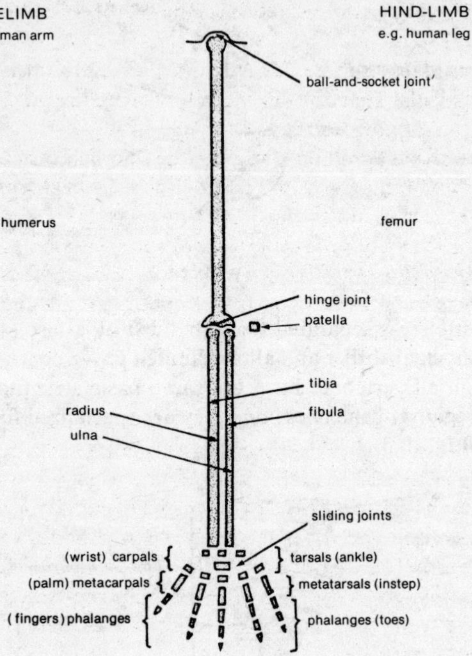

Fig. 10.4. *Basic plan of the limb skeleton* (*the pentadactyl limb*)

(i) The 7 **cervical** vertebrae support the neck while providing some flexibility of movement. The first vertebra, the **atlas**, supports the skull and allows nodding ('yes') movements. The second vertebra, the **axis**, projects into the atlas and allows swivelling ('no') movements. The other vertebrae have small processes and also permit twisting of the neck. The vertebrae in this region have large holes parallel to the neural canal which allow blood vessels through to reach the brain.
(ii) The 12 **thoracic** vertebrae support the ribs and allow bending and rotation of the trunk. They have long spines for attachment of shoulder and back muscles. The transverse processes are reduced because of the presence of ribs.
(iii) The 5 **lumbar** vertebrae are thick-set with large processes and wide spines for attachment of muscles. These vertebrae provide the major support for the trunk and are subject to great stress. They allow bending and rotation of the trunk.
(iv) The 5 **sacral** vertebrae are fused together for strength and form part

of the pelvis. They transmit the weight of the body to the pelvis and hips.
(v) The 4 **caudal** vertebrae in humans are fused to form the **coccyx** and have no particular function. In the rabbit there are 16 caudal vertebrae; they form the support for the tail.

The **ribs** and **sternum**, together with the thoracic vertebrae, enclose the thoracic cavity and protect the heart and lungs. Intercostal muscles are attached to the ribs. When they contract and relax they alter the volume of the thoracic cavity producing ventilation of the lungs (see page 104). The ribs are connected to the sternum by cartilage.

2. **The appendicular skeleton** (girdles and limbs): the **girdles** are box-like arrangements of bone which connect the limbs to the axial skeleton.

The **pectoral** (shoulder) **girdle** is connected to the vertebral column by muscles and tendons. The girdle consists of a pair of **scapulas** (shoulder blades) and a pair of **clavicles** (collar bones). The forelimbs (in humans, the arms) are jointed to the scapula and are attached to the pectoral girdle by muscles and tendons. This arrangement allows great mobility of the forelimbs coupled with shock-absorbing ability.

The **pelvic girdle** (pelvis) is made up of fused bones which in turn are fused to the sacral vertebrae (see above). The two hind-limbs (in humans, the legs) are jointed to the pelvis. The pelvis is rigid and strong to enable the upthrust from the hind-limbs to be transmitted effectively to the vertebral column.

The **limbs** are based on the pentadactyl ('five finger') plan typical of higher vertebrates (see Fig. 10.4). The basic plan is modified according to:

(a) whether it is a fore- or hind-limb
(b) the method of locomotion, e.g. swimming, walking or flying.

Having looked at the mammalian skeleton try the following question:

Q. (a) Give three functions of the skeleton of a mammal.
(b) What are the major chemical constituents of bone?
(c) Name the principal regions of the spine and for each region give:
(i) one special function;
(ii) one adaptation to performing that function.

Joints

A joint is the place where two bones meet. Joints can be broadly classified as movable, slightly movable or immovable.

1. **Immovable** joints (sutures) join bones together in the cranium and pelvic girdle. Where bones meet they are connected by the fibrous protein **collagen**.

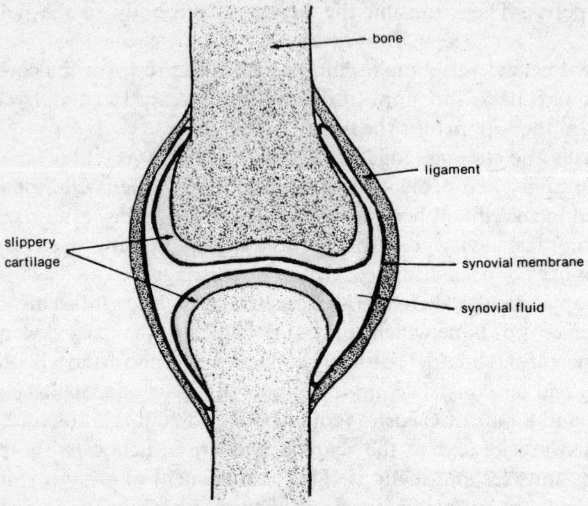

Fig. 10.5. *Structure of a synovial joint*

2. **Slightly movable** (cartilaginous) joints are found where bones are separated by cartilage and tightly bound together by ligaments, e.g. the joints between vertebrae.
3. **Movable** (synovial) joints are associated with the limbs. Where the bones come together they are covered with shiny, slippery cartilage. Between the cartilages is **synovial** fluid which allows almost frictionless movement between the bones. The fluid is secreted by the synovial membrane which surrounds the whole joint (see Fig. 10.5). Ligaments hold the bones together. The arrangement shown in Fig. 10.5 applies to the ball-and-socket and hinge joints described below.

Movable joints fall into one of three groups:
(a) **Ball-and-socket joints** (also called universal joints): these allow movement in all planes; up and down, side to side, and rotation. These joints are found between the pelvic girdle and femur (hipbone) and between the scapula (shoulder blade) and humerus (upper arm).
(b) **Hinge joints:** these allow movement in one plane only, like the hinge of a door. They are found at the knee and elbow.
(c) **Sliding joints:** these are found where the flat surface of one bone glides across the surface of another. Such joints give flexibility to the wrist and ankle.

10.3. *Muscle*

Bones are moved relative to each other by contraction of muscles. There are three types of muscle, but only one type is associated with the skeleton:

1. **Striped (skeletal) muscle:**

(a) It is attached to the skeleton and provides the force required for body movement and locomotion.

(b) It contracts powerfully when stimulated by nerve impulses (page 164) but soon fatigues (tires).

(c) Under the microscope the elongated muscle fibres have a striped appearance (hence the name). There is no clear-cut separation between one cell and the next (see Fig. 10.6).

(d) Muscle contraction is normally under conscious control, hence these muscles are sometimes called 'voluntary' muscles.

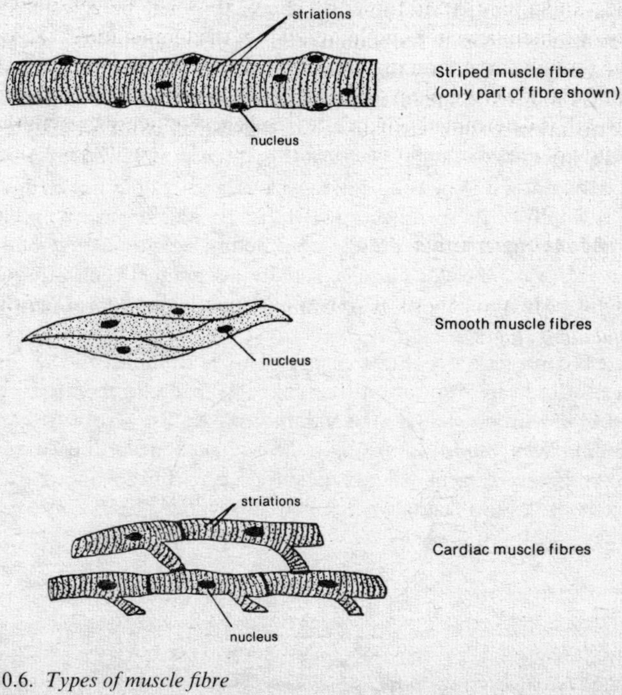

Fig. 10.6. *Types of muscle fibre*

The other two types of muscle are:

2. **Smooth (visceral) muscle:**

(a) This is found in the walls of internal organs such as the gut and bladder.

(b) When stimulated by nerve impulses, it contracts less powerfully than striped muscle, but it is longer acting, i.e. it does not fatigue so easily.

(c) The cells are unstriped and are clearly separate from each other (Fig. 10.6).

(d) Muscle contraction is not under conscious control, hence these muscles are sometimes called 'involuntary' muscles.

3. **Cardiac (heart) muscle:**

(a) This is found in the walls of the heart.

(b) It is **myogenic** (contracts of its own accord *without* nervous stimulation). NOTE. The heart is under nervous control, but this is to speed up or slow down heartbeat, not initiate it (see page 127).

(c) Cardiac muscle is unfatiguing (if your heart stopped beating you would soon die).

(d) The cells are striped, individual cells are distinguishable and the cells (fibres) are branched (Fig. 10.6).

Muscles are made up of thousands of **muscle fibres** which have the ability to contract (shorten) and relax. When they contract, they exert a force, In the case of skeletal muscle this force usually moves one bone relative to another. When skeletal muscle relaxes it does not return to its original length of its own accord. It must be pulled out by gravity or, more usually, by another muscle contracting against it and pulling it out. In this way, skeletal muscles usually act against each other in **antagonistic pairs** (see Fig. 10.7). When one muscle contracts the other relaxes.

Muscle contraction requires energy. This is obtained in the form of ATP, produced by respiration (see page 92). Muscles therefore require a good blood supply delivering food and oxygen for aerobic respiration (page 92). This blood supply can be adjusted according to muscle need. At times of peak energy demand, e.g. during heavy exercise, aerobic respiration is supplemented by anaerobic respiration (see page 92).

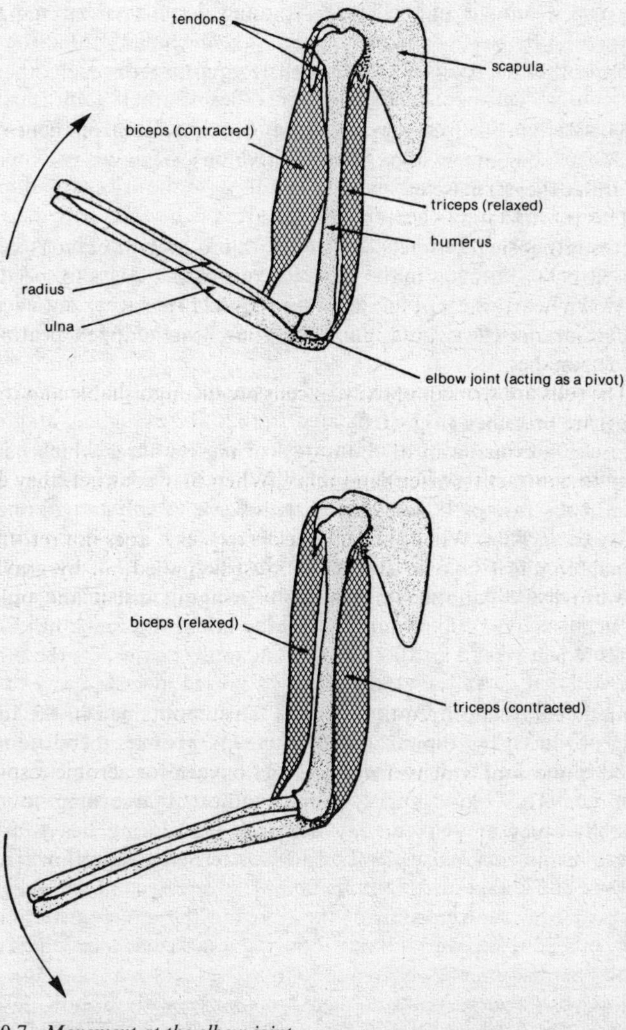

Fig. 10.7. *Movement at the elbow joint*

10.4. *A Limb Joint*

You need to be familiar with one *named* limb joint; how it works and how bones, muscles, tendons and nerves bring about movement at the joint. We will choose the elbow joint as our example.

The elbow joint is of the hinge type and permits an up and down movement of the lower arm. However, unlike the hinge joint at the knee, it also allows some rotation (you can twist your forearm).

The muscle which contracts and bends (flexes) a limb joint is called a **flexor muscle**; in this case it is the **biceps** (see Fig. 10.7). The muscle on the opposite side of the limb joint, and which straightens (extends) the limb, is called the **extensor muscle**, in this case the triceps. Notice the biceps muscle is larger than the triceps. This is because it is the major load-bearing muscle and acts against the pull of gravity. The triceps, on the other hand, is acting in the direction of the pull of gravity. With the palm of the hand facing upwards, contraction of the biceps muscle raises the forearm; contraction of the triceps muscle and relaxation of the biceps lowers the forearm.

To produce effective movement it is essential that the contraction of various sets of muscles is coordinated so that, for example, antagonistic muscles do not contract at the same time. Each muscle has both a motor and sensory nerve supply. Nerve impulses from the motor area of the cerebral cortex (page 170) stimulate the muscle to contract. Within the muscle are stretch receptors which, when stretched, send nerve impulses along sensory neurones to the spinal cord and brain. These receptors supply information about the position of the limbs and enable a pattern of muscular activity to be computed by the brain, so producing effective movement.

Muscles that move bones are usually attached in such a way that the bones act as levers with low mechanical advantage. Fig. 10.7 makes this clear. The muscles of the upper arm only contract over a short distance but, being attached near to the joint, they magnify the movement at the end of the limb. The biceps muscle may contract only 8 or 9 cm but the hand will move about 60 cm.

The skeletal muscles of the body are rarely in the state where one muscle of an antagonistic pair is completely relaxed. Even when the body is at rest, a proportion of the fibres in the antagonistic muscles remain in a state of contraction producing a **muscular tone** which holds the body in position.

Using your course notes to help you, answer the following question:

Q. Make a labelled diagram of a named *limb of a mammal to show the arrangements of the bones, tendons, muscles and nerves which bring about movement at a joint. Include details of the joint itself.*

10.5. *Types of Locomotion*

On certain syllabuses you are required to know a range of methods of locomotion in vertebrates, including swimming, crawling, leaping, walking, running and flapping flight. Check your syllabus requirements. We will examine two methods of locomotion which are common to most syllabuses: swimming and flapping flight.

Swimming in a bony fish, e.g. herring

The body of the fish is well adapted for movement in water:
1. The streamlined body form, backward-pointing scales and mucous covering provide little resistance to movement through water.
2. Various **fins** are present which provide
(a) propulsion
(b) stability.
3. The flexible body and vertebral column allow swimming movements.
4. Muscle blocks (myotomes) on either side of the vertebral column provide the power for swimming.
5. An air bladder may be present which provides buoyancy and makes the fish effectively 'weightless' in water.

Propulsion: the muscular tail has a large surface area which provides the force which propels the fish foward. This is achieved by alternate contraction and relaxation of muscle blocks (myotomes) lying on opposite sides of the vertebral column (Fig. 10.8b). The tail and body exert a sideways and backwards thrust against the water and the resistance provided by the water pushes the fish sideways and forwards. The sideways forces cancel out and the overall movement is therefore forwards (see Fig. 10.8a).

Stability: the fins give the fish a very precise steering capability and provide stability. The vertical fins, the dorsal, ventral and tail (caudal) fin prevent the fish **rolling** and **yawing** (see Fig. 10.9). The paired pectoral and pelvic fins prevent **pitching** and act as brakes to slow down or stop the fish (Fig. 10.9).

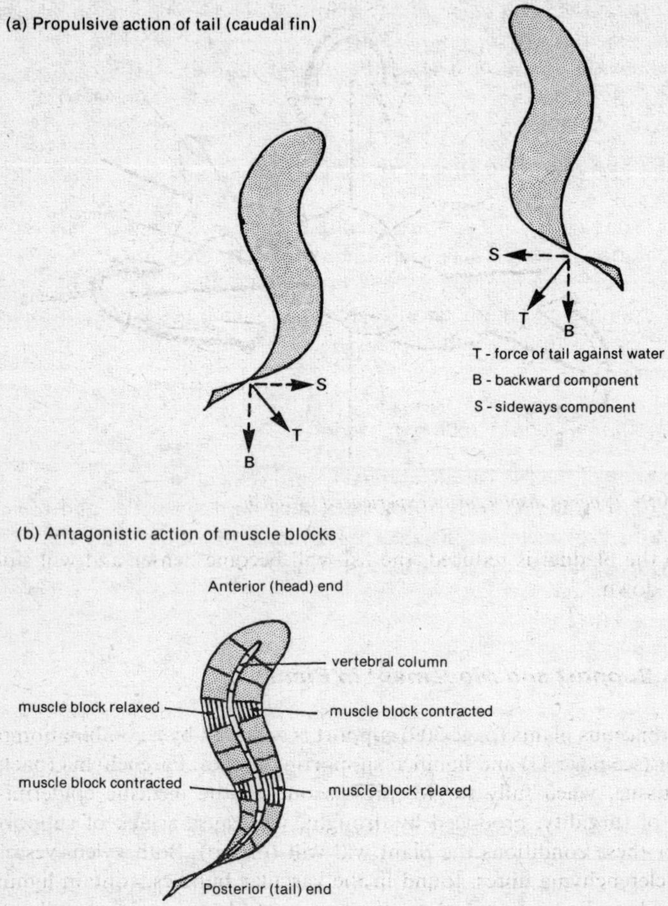

Fig. 10.8. *Propulsion in fish*

Buoyancy: most bony fish are almost weightless in water and so, unlike birds, do not have to expend much energy in overcoming the force of gravity. They can use most of their energy in moving through the water. This weightlessness (buoyancy) is achieved by the possession of an air-filled space within the body, the **swim bladder**. The amount of gas in the bladder can be regulated. If the swim bladder is filled with air, the fish becomes less dense and will float higher up in the water. If the amount of

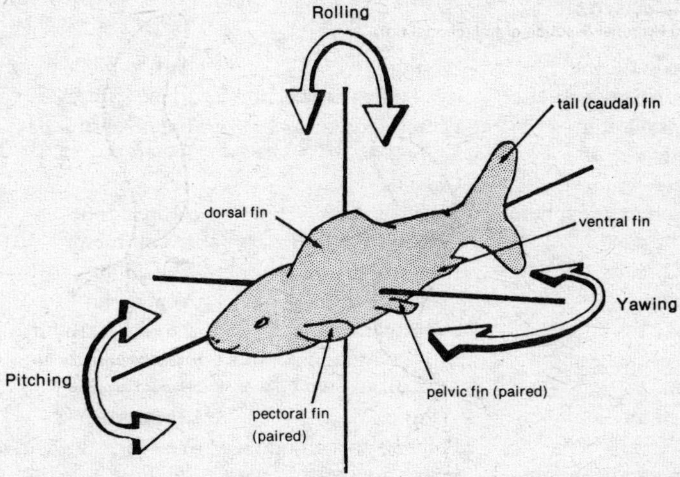

Fig. 10.9. *Types of displacement experienced by a fish*

air in the bladder is reduced, the fish will become denser and will sink lower down.

10.6. *Support and Movement in Plants*

In herbaceous plants (page 309) support is achieved by a combination of turgor (see page 43) and lignified supporting tissues. Parenchyma (packing) tissue, when fully turgid, presses out on the inelastic epidermis. Lack of turgidity, produced by drought, will cause a lack of support. Under these conditions the plant will wilt (droop). Both xylem vessels and sclerenchyma fibres, found in the vascular bundles, contain lignin, and add mechanical strength to the stem, roots, leaves and leaf stalks.

In the stem of woody perennials (see page 309) support is entirely achieved by lignified tissues. Each year a new layer of xylem is added to the stem. This layer ceases to be functional in transporting water and salts, and takes on a purely supportive role. The heartwood in older trees is composed of xylem vessels and sclerenchyma fibres used solely for support.

Plants, in general, show slow movements as a result of growth and faster movements as a result of changes in turgidity. For details of growth movements see pages 180–87.

Key Words

Locomotion	Sternum	Antagonistic muscles
Hydrostatic skeleton	Appendicular skeleton	Flexor muscle
Exoskeleton	Pectoral girdle	Extensor muscle
Cuticle	Scapula	Biceps
Bone	Clavicle	Triceps
Calcium phosphate	Pelvic girdle	Muscle tone
Collagen	Pentadactyl limb	Myotomes
Ligaments	Immovable joints (sutures)	Dorsal fin
Tendons	Slightly movable (cartilaginous) joints	Ventral fin
Muscles	Movable (synovial) joints	Caudal (tail) fin
Axial skeleton	Ball and socket joints	Pectoral fins
Skull	Hinge joints	Pelvic fins
Cranium	Sliding joints	Rolling
Jaws	Striped (skeletal) muscle	Yawing
Vertebral column	Smooth (visceral) muscle	Pitching
Cervical vertebrae	Cardiac (heart) muscle	Swim bladder
Thoracic vertebrae		Turgor
Lumbar vertebrae		Xylem vessels
Sacral vertebrae		Sclerenchyma
Caudal vertebrae		Lignin
Ribs		Heartwood

Question

1. (*a*) *List the names of the types of joints found in the skeleton and give* one *example of where each type might be found.*
(*b*) *Make a large, labelled diagram to show the chief muscles and bones involved in movement at the human elbow.*
(*c*) *Briefly explain why:*
(*i*) *the muscles shown in your diagram are often described as being 'antagonistic'*
(*ii*) *each muscle has a nerve and blood supply.*

11. Reproduction

Definition: Reproduction *is the production of new individuals* (*offspring*) *derived from existing individuals* (*parents*).

Reproduction is a characteristic of all living organisms. There are two forms of reproduction, **asexual** and **sexual**.

11.1. *Asexual and Sexual Reproduction*

Asexual reproduction has the following features:

1. New individuals (offspring) are formed from *one* parent.
2. All offspring are genetically identical to each other and to the parent.
3. Asexual reproduction involves the type of cell division called **mitosis** (page 252).

Asexual reproduction has two main advantages:

1. Under favourable conditions, large numbers of offspring can be produced in a short time.
2. The offspring, being genetically identical to the parent, are normally able to survive under the same set of conditions.

Sexual reproduction has the following features:

1. A new individual is formed from *two* cells (**gametes**) which fuse to form a cell called a **zygote**. The two gametes usually come from two different individuals of the same species.
2. The offspring are genetically different to the parents and usually to each other. This generates variation within the species (see page 260).
3. The type of cell division called **meiosis** (page 252) gives rise to the gametes.

Sexual reproduction has one very important advantage: the offspring, being genetically different to the parents, are able to survive under different sets of conditions. This enables a species to withstand changes in an existing environment and to move into new environments.

Now try the following questions:

Q. Distinguish carefully between sexual and asexual reproduction.

Q. Give one *advantage of asexual reproduction and* one *advantage of sexual reproduction.*

11.2. *Types of Asexual Reproduction*

Asexual reproduction is common throughout the plant kingdom but only occurs in simpler animals. Some of the more common methods are:

1. **Binary fission:** division of a unicellular organism into two identical individuals which then separate, e.g. in bacteria (page 294), *Chlamydomonas* (page 297) and *Amoeba* (page 315).
2. **Spore formation:** cells divide by mitosis to produce spores (single cells enclosed in a resistant coat). These disperse from the parent and under appropriate conditions each spore can develop into a new individual, e.g. in *Mucor* (page 301).
3. **Budding:** an outgrowth (bud) develops from the body wall. The bud grows to form a daughter organism which detaches to become independent, e.g. in *Hydra*.
4. **Vegetative propagation:** (see Table 11.1 and Fig. 11.1): this is a common method of asexual reproduction in flowering plants. It occurs when part of the parent plant becomes detached and develops into a new individual. This type of reproduction is often associated with storage organs. In herbaceous perennials (page 309) these storage organs may also act as organs of perennation.

Vegetative propagation has a number of advantages:

(a) The offspring, being genetically identical to the parent, are able to survive under the same set of conditions.

(b) The hazards involved in seed germination are avoided, e.g. the supply of food and water is provided by the storage organ or parent plant until the new plant is well established.

(c) Vegetative propagation tends to produce clumps of plants. This reduces the room for competition from other types of plants.

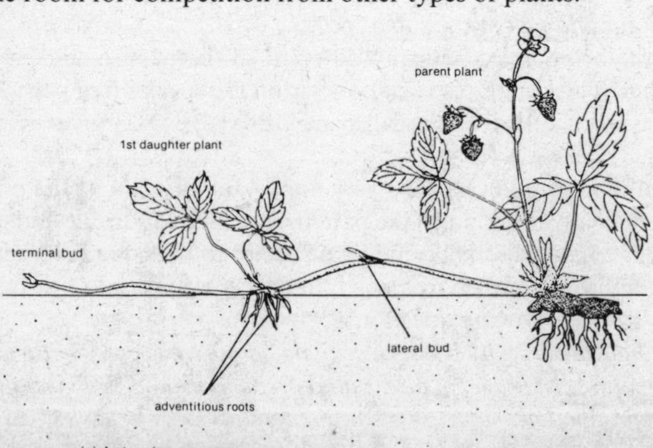

(a) A strawberry runner

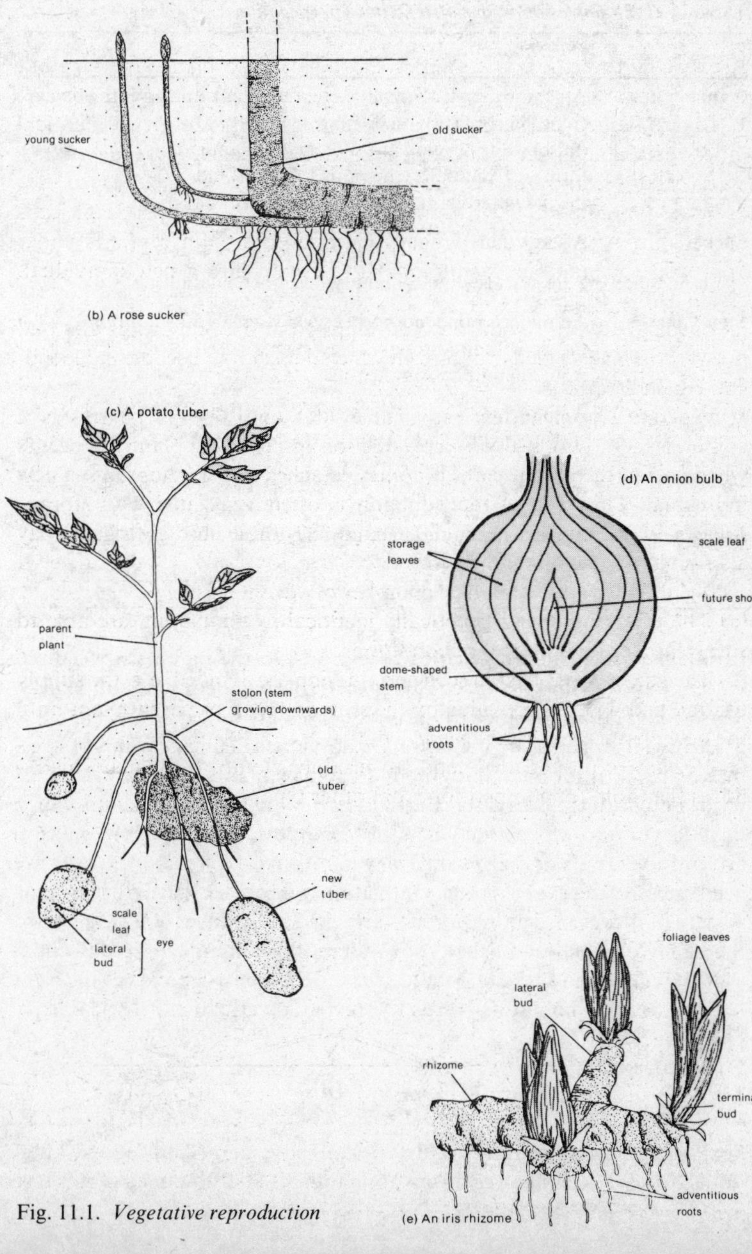

Fig. 11.1. *Vegetative reproduction*

Table 11.1. *Examples of organs of vegetative propagation*

Organ	*Description*	*Examples*
Runner	An overground, horizontal stem. If a bud comes in contact with damp soil, adventitious (false) roots develop and a daughter plant is formed. Eventually, the runner shrivels and the daughter plant becomes independent	Strawberry
Sucker	A short underground runner	Rose
The following are storage organs and organs of perennation:		
Stem tuber	An underground lateral stem which swells at the tip with food reserve. New plants arise from lateral buds on the tuber	Potato
Bulb	An underground stem growing vertically and surrounded by fleshy leaves containing food reserves. New plants arise from lateral buds at the base of fleshy leaves	Onion
Rhizome	An underground stem growing horizontally and swollen with food reserves. New plants arise from lateral buds on the rhizome	Iris

(d) As the offspring are genetically identical to the parents, specialized varieties and quality plants can be grown (propagated) without change. This is of great importance to the commercial grower.

5. **Artificial propagation:** humans have developed artificial means of vegetative propagation by making buds form new plants. The two methods used are cuttings and grafts. **Cuttings** are short lengths of shoot which bear leaves and are planted in moist compost. They quickly develop adventitious roots and grow into new plants (e.g., *Geranium*). **Grafts** are used most extensively in the cultivation of rose and fruit trees. The variety of flower or fruit required is attached in the form of a twig, called the **scion**, to a well-established root system, called the **stock**. It is essential that the cambium of the scion and stock are in contact and that the joint between them is bound with tape to prevent infection and frost damage.

11.3. *Types of Sexual Reproduction*

Sexual reproduction occurs widely throughout plant and animal kingdoms, from unicellular organisms to mammals and flowering plants (see

Table 11.2. *The occurrence of asexual and sexual reproduction in plant and animal kingdoms*

Plants	*Asexual*	*Sexual*	*Two Sexes*	*Hermaphrodite*
Chlamydomonas	Yes	Yes	Strains	No
Spirogyra	Yes	Yes	Strains	No
Mucor and Phytophthora	Yes	Yes	Strains	No
Fern	Yes	Yes	Sometimes	Usually
Herbaceous plant	Yes	Yes	Sometimes	Usually
Deciduous tree	Yes	Yes	Sometimes	Usually
Animals				
Amoeba	Yes	No	—	—
Earthworm	No	Yes	No	Yes
Butterfly Bee	No	Yes	Yes	No
Fish	No	Yes	Yes	No
Frog	No	Yes	Yes	No
Bird	No	Yes	Yes	No
Mammal	No	Yes	Yes	No

Table 11.2). Even bacteria carry out a form of sexual reproduction called conjugation.

Remember the main points about sexual reproduction:

1. It involves fusion of two gametes to form a zygote.
2. The production of gametes involves meiosis.
3. The offspring are genetically different to the parents.

The body cells of most organisms contain two sets of chromosomes, one set derived from the male parent and one from the female parent. This is called the **diploid** number of chromosomes (2n).

Definition: The diploid *number (2n) is the number of chromosomes in a body cell. It consists of two sets of similar chromosomes.*

We know that in sexual reproduction a new individual is formed by the fusion of two cells called gametes. If these cells had the normal chromosome complement of body cells, the zygote which results would have twice the diploid number of chromosomes (4n). This is not the case. The gametes contain only *half* the chromosome number of body cells, so that when they fuse they restore the normal chromosome number. The gametes have the **haploid** number of chromosomes (n).

Definition: The haploid *number (n) is the number of chromosomes in a gamete. It consists of one set of chromosomes.*

When male and female gametes fuse at fertilization, the diploid chromosome number is restored:

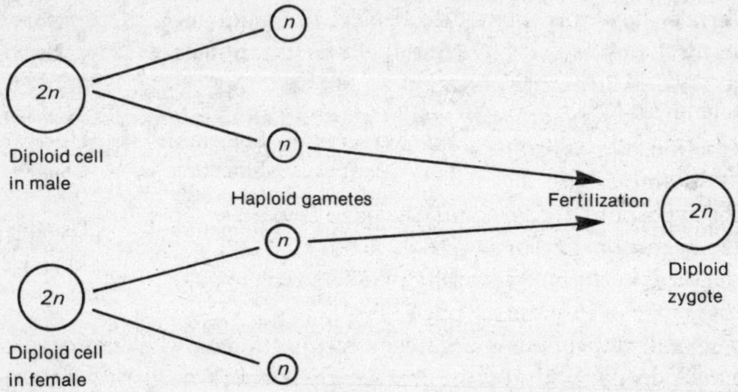

Definition: Fertilization *is the fusion of male and female gamete nuclei to form a zygote.*

In simpler organisms such as *Spirogyra* (page 299) and *Mucor* (page 301) the gametes are not produced by two distinct sexes, but rather there are a number of strains. Sexual reproduction occurs between different strains.

In more complex organisms two different sexes, male and female, are recognized. Certain parts of the body, the **reproductive organs** or **gonads**, have become specialized for the production of gametes. The **egg** is the gamete produced by the female. It is larger than the male gamete, is immobile and contains a food reserve. The **sperm**, the gamete produced by the male, tends to be much smaller and moves to meet the female gamete. In animals and lower plants the sperm is motile (moves by itself), whereas in flowering plants the male gamete is carried by external agents such as wind or insects.

In some organisms called **hermaphrodites**, individuals have both male and female reproductive organs and produce both kinds of gamete (see Table 11.2).

Definition: A hermaphrodite *is an individual which has both male and female reproductive organs and is able to produce both male and female gametes.*

Being hermaphrodite has two major advantages:

1. An individual can mate with any other adult individual of the same species and does not have to wait for an individual of the opposite sex.

2. In the absence of a partner, a hermaphrodite can potentially self-fertilize. However, in most hermaphrodite organisms, self-fertilization is prevented since this rather defeats the main object of sexual reproduction – to generate variation.

Certain flowering plants do self-fertilize, but most hermaphrodites have mechanisms for preventing self-fertilization (see the earthworm, page 317 and flowering plants, page 308).

In a number of groups, sexual reproduction is associated with the production of a resistant stage which can survive under adverse conditions. Examples are:

1. the zygospore of *Spirogyra* (see page 299)
2. the zygospore of *Mucor* (see page 301)
3. the seeds of annual flowering plants (see page 308)
4. the pupae of insects (see page 319).

In sexually reproducing organisms fertilization may be **external**, occurring outside the body of the female, or **internal**, occurring within the female's body.

External fertilization is common in aquatic (water-living) organisms where water provides the medium in which sperm can swim to eggs. It also prevents gametes, zygotes and developing offspring from drying out.

Internal fertilization is a feature of terrestrial (land-living) organisms where there is little free water and little opportunity for male gametes to swim freely.

In both higher plants and vertebrates there is a trend towards internal fertilization in terrestrial (land) environments. Ferns (page 305) and mosses (page 304) live on land, have external fertilization, and are dependent on water for transference of the male gamete to the female reproductive organs (page 307). Flowering plants have internal fertilization and are *not* dependent on water for transferring of the male gamete (see page 215).

Among the vertebrates (page 323), fish (which are obviously aquatic) have external fertilization, and so do amphibia (they return to water to breed). Birds, reptiles and mammals have internal fertilization.

In fish and amphibia large numbers of eggs are laid and fertilization rates are relatively low. The fertilized eggs are usually given little or no protection, there is little parental care, and the young hatch out at a relatively undeveloped stage. Mortality rates are high, hence the large number of eggs produced.

In terrestrial vertebrates, internal fertilization replaces external fertilization, a suitable aquatic medium being provided within the body. There is also the problem of offspring drying out and different strategies have been developed to prevent this happening. In birds, the embryo is enclosed in a chalky shell together with a large food store. The eggs are usually incubated (kept at a constant temperature) by the parents and the young birds after hatching are fed until they can fly and fend for themselves. In mammals, the developing offspring are retained within the body of the female and the offspring are born in a relatively advanced stage of development. After birth the offspring receive parental care.

Internal fertilization is associated with a high degree of protection for developing offspring, together with parental care after birth or hatching. The greater the energy expended in protecting and caring for the young, the fewer the number of offspring produced, and the lower the mortality rate. This pattern is evident in Table 11.3.

Table 11.3. *Comparison of reproduction in fish, birds and mammals*

Fish, e.g. herring (see page 323)	*Bird*, e.g. pigeon (see page 327)	*Mammal*, e.g. rabbit (see page 329) and human (see page 223)
1. Mating usually seasonal	Mating usually seasonal	Mating usually seasonal (not in humans)
2. External fertilization	Internal fertilization (in the oviduct)	Internal fertilization (in the oviduct)
3. Thousands of eggs produced at one time	Few eggs produced at one time	Few eggs produced at one time (in humans, one)
4. Small amount of yolk in the egg	Egg contains large amount of yolk	Little or no yolk with egg (embryo supplied with food via placenta)
5. Development of embryo usually external and unprotected	Development of embryo external but within egg shell, incubated and protected by parents	Development of embryo protected within uterus of mother
6. Young hatch from egg, but usually unprotected	Young hatch from egg, but protected	Mother gives birth to live young
7. No parental care (usually)	Parental care. Young are protected, fed and taught	Parental care. Young are protected, suckled and taught
8. Chance of completing development low	Chance of completing development high	Chance of completing development high

11.4. *Sexual Reproduction in Flowering Plants*

In a flowering plant, the **flower** is the organ of sexual reproduction. Angiosperms (page 308) do not exhibit the alternation of generations shown by mosses (page 304) and ferns (page 305). In reality, however, the gametophyte and sporophyte generations are present, but in a disguised form. The pollen and ova represent the male and female gametophytes, while the flower and rest of the plant represent the sporophyte generation.

Mosses and ferns are dependent on water for the transport of the male gamete to the female gamete and for successful fertilization. Flowering plants have overcome this need. In effect, fertilization takes place internally, within the ovule of the flower. The male gamete is contained within the pollen grain which is resistant to drying out. It reaches a female sex organ on the wind or is carried by an insect.

Spores are the dispersal phase in mosses and ferns whereas in flowering plants the seed and fruit are.

Sexual reproduction in flowering plants takes the following pattern:

Flower formation ←
↓
Pollination
↓
Fertilization
↓
Seed and fruit formation
↓
Dispersal
↓
Germination
↓
Growth to adult plant → (back to Flower formation)

Flower structure

Flowers are very varied in shape and structure but all conform to a general plan. This is best shown by looking at a simple flower such as the buttercup. You are required to know the structure, and the functions of the parts, of *one* insect-pollinated flower (see Fig. 11.2).

The parts of the flower are attached to the **receptacle**, the tip of the flower stalk. The parts are arranged in four rings, or **whorls**.

The outermost whorl is called the **calyx**. It is made up of five **sepals** –

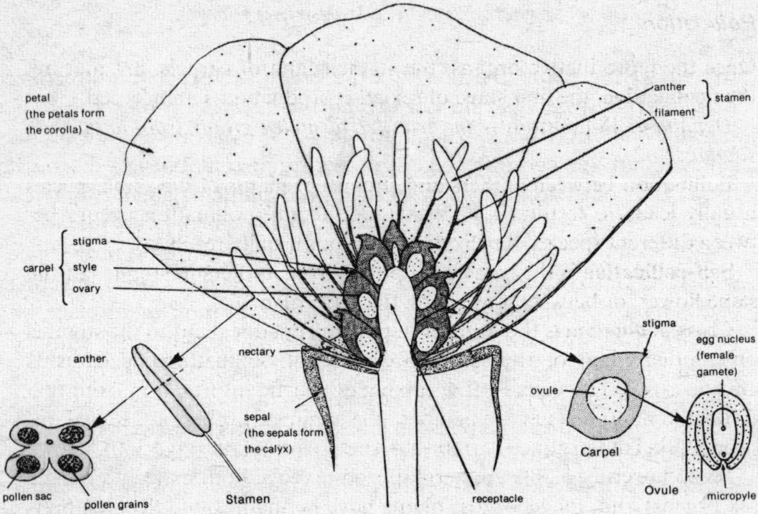

Fig. 11.2. *Structure of a buttercup flower* (*insect-pollinated*)

leaf-like structures which protect the developing flower bud. They fold back when the flower blooms.

The next whorl is called the **corolla**. It is made up of five **petals**. In insect-pollinated flowers such as the buttercup the petals are brightly coloured to attract insects. Some also have scented petals with a **nectary** at the base which produces sugary nectar. This also attracts insects.

Inside the corolla is the **androecium**. This is the male part of the flower and consists of a number of **stamens**. Each stamen is made up of a **filament** supporting an **anther**. Inside the anthers are pollen sacs in which the **pollen grains** (containing **male gametes**) develop.

The innermost whorl is the **gynoecium**. It is the female part of the flower and is made up of the female reproductive organs, the **carpels**. Some plants, such as the buttercup, have separate carpels, whereas others have a number of carpels fused together, as in the primrose. Each carpel consists of a wide, hollow base called the **ovary**, above which is a narrow region, the **style**. This ends in a sticky platform, the **stigma**, which receives pollen grains from the same or another plant during pollination.

Within the ovary are varying numbers of **ovules**. The buttercup has one ovule inside each ovary (see Fig. 11.2). At the centre of each ovule is a large cell called the **embryo sac**, and this contains several nuclei, one of which is the female gamete, or **egg nucleus**. It is this which is fertilized by a male nucleus from a pollen grain.

Pollination

Once the reproductive organs, the stamens and/or carpels, are mature, then **pollination**, the first stage of sexual reproduction, can proceed.

Definition: Pollination *is the transfer of pollen grains from anthers to stigmas.*

Pollination between anthers and stigmas of plants of the same species usually leads to fertilization. When 'accidental' pollination occurs between different species, fertilization does not usually result.

Self-pollination is the transfer of pollen from anthers to stigmas of the same flower, or between flowers on the same plant.

Cross-pollination is the transfer of pollen from one plant to the stigmas on another plant of the same species. Most pollination mechanisms ensure cross-pollination so that mixing of genetic information from two different plants occurs (remember, the main advantage of sexual reproduction is to produce variation in the offspring, see page 260).

Most flowering plants are hermaphrodite (have both male and female sex organs) and, indeed, most plants have hermaphrodite flowers. Self-fertilization is usually prevented by the male flower parts, the stamens, maturing before or after the female parts, the carpels. When the stamens mature before the carpels, e.g. in the buttercup, it is called **protandry**. Much less common is **protogyny**, where the carpels mature first, e.g. in the bluebell. In some species there are separate male and female flowers, e.g. in the willow tree, and in others separate male and female plants, e.g. in the cucumber plant.

In most flowers cross-pollination involves one of two agents, insects or wind, to transfer pollen from one flower to another. Which particular

Table 11.4. *Comparison of features of insect- and wind-pollinated flowers*

Insect-pollinated	*Wind-pollinated*
1. Large, brightly coloured petals	Small petals, not brightly coloured
2. May have nectar	No nectar
3. May be scented	Not scented
4. Small anthers (inside flower)	Large anthers (outside flower) which release pollen into the wind
5. Large spiny, sticky pollen grains produced in small number (to adhere to bodies of insects)	Small, light and smooth pollen grains produced in large quantities (only a very few will reach a stigma)
6. Sticky stigma inside the flower to obtain pollen from insect	Large feathery stigma(s) hanging outside the flower to trap pollen on the wind

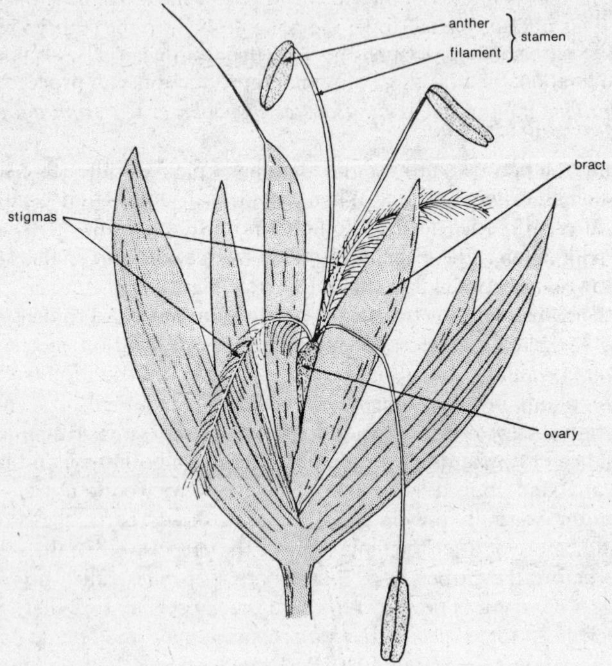

Fig. 11.3. *An oat flower (wind-pollinated)*

agent is used can be determined by the appearance of the flower. The differences between wind- and insect-pollinated flowers are a common feature in exams (see Table 11.4). An oat (Fig. 11.3) is an example of a wind-pollinated plant.

Fertilization

Once a pollen grain has arrived at the stigma of a flower of the same species, the male gamete has yet to meet the female gamete. Fertilization only occurs when the two gametes fuse.

The path between the male and female gamete nuclei is cleared by a remarkable process. Sugary secretions on the stigma stimulate the pollen grain to produce a pollen tube. The tube grows down towards the ovule (probably guided by chemicals) and then enters the ovule by a tiny hole, the **micropyle**. The tip of the pollen tube opens forming a clear pathway through which the male gamete reaches the female gamete (see Fig.

11.4).The male gamete is a nucleus from one of the cells inside the pollen grain. This nucleus passes down the pollen tube into the embryo sac and brings about fertilization by fusing with the egg nucleus.

Fruit and seed formation

After fertilization the ovary undergoes changes to produce the **fruit**, the structure which contains and protects the **seeds**. The ovule develops into the seed and the ovary wall becomes the fruit. The other parts of the flower, the stamens, petals and sepals, are no longer needed: they shrivel up and drop off.

Inside the seed, the zygote divides many times by mitosis to become an embryo plant. This consists of:

1. a young root (the **radicle**)
2. a young shoot (the **plumule**)
3. one or two seed leaves (**cotyledons**). Monocotyledonous plants have one seed leaf, dicotyledonous plants have two (see page 308).

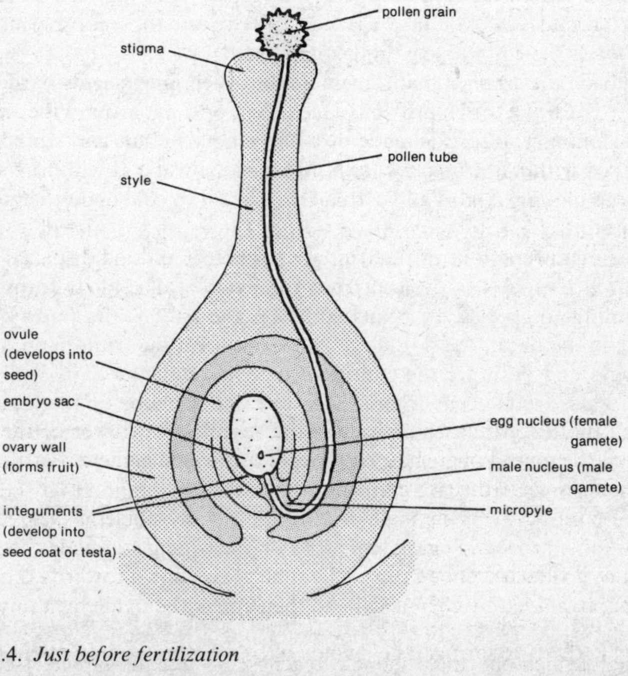

Fig. 11.4. *Just before fertilization*

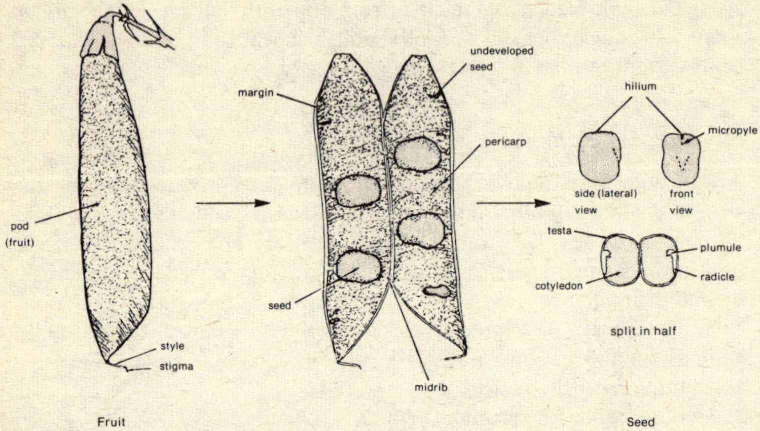

Fig. 11.5. *Fruit and seed of a pea*

Stored within the seed are food reserves, either in the seed leaves, e.g. in the broad bean, or in a mass of cells called the endosperm which surrounds the embryo, e.g. in the maize seed.

Outside the embryo plant, membranes called **integuments** harden and lignify to form a tough protective seed coat, or **testa**, around the embryo.

In summary, a seed is made up of an embryo plant and stored food, enclosed within a protective testa. Most of the water is withdrawn from the seed making it hard and extremely resistant to cold and other adverse conditions.

The main functions of the fruit are to protect and aid dispersal of the seeds. A **true fruit** is derived from the ovary wall and its form varies according to species. In beans and peas the fruit is the pod (see Fig. 11.5); in the plum, the skin, flesh and stone form the fruit; in an orange, the rind and flesh are the fruit, and in a hazelnut the hard shell is the fruit. **False fruits** occur where the ovary wall remains undeveloped. Instead, the receptacle enlarges to form the fleshy part attached to or enclosing the seeds, e.g. in apples, pears and strawberries.

The form the fruit takes is usually linked to the method of dispersal.

Fruit and seed dispersal

As plants are generally fixed, they must employ effective methods to disperse their offspring. This avoids overcrowding and competition for

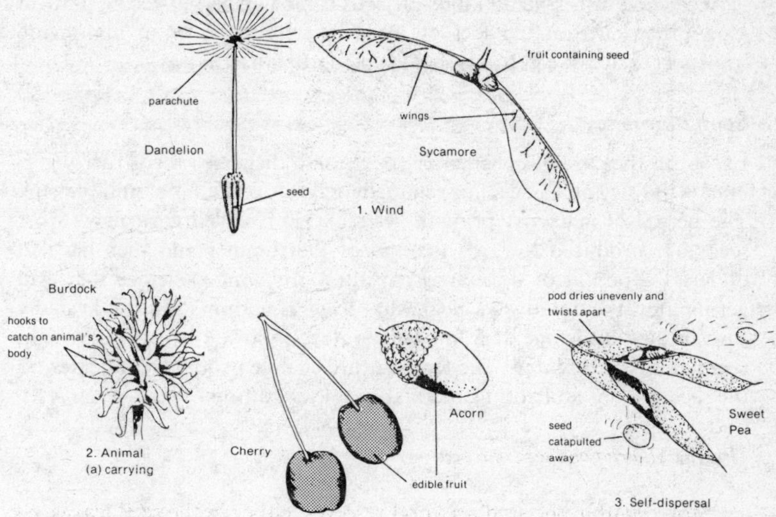

Fig. 11.6. *Methods of seed dispersal*

space, light and water, and also enables the spread of plants into new and different environments. The three main types of dispersal are:

1. **Wind dispersal:** the fruits of certain trees such as the ash and sycamore (see Fig. 11.6) develop wing-like structures. The fruits of certain herbaceous plants such as dandelion (Fig. 11.6) and willow herb develop into parachutes. In both types of fruit, the development of a large surface area relative to volume slows the fall of the fruit to the ground and produces dispersal over a wider area.

2. **Animal dispersal:** there are two main types:

(a) The fruit or seed may develop hooks which catch on the fur of animals, e.g. the fruit of the burdock (Fig. 11.6).

(b) The fruit may become fleshy (succulent), brightly coloured and sweet tasting. Small seeds, e.g. those of the blackberry or strawberry, can be ingested and pass through the animal's digestive system relatively unharmed. In such cases, the seed may be carried some distance before being deposited on the ground along with a convenient supply of fertilizer in the faeces. In other fruits, e.g. the cherry, the seeds in their hard coats are discarded after the soft fruit has been eaten.

3. **Self-dispersal:** the fruits of several plants such as the sweet pea, gorse, lupin and geranium develop into a pod or similar structure which, when it dries, catapults seeds away from the parent plant (see Fig. 11.6).

Accidental dispersal: all kinds of seeds can be dispersed accidentally in mud picked up on the feet of birds or mammals, or in man-made transport systems, e.g. trains, boats and aeroplanes.

Seed dormancy

Except in the case of ephemerals (page 308), dispersed seeds rarely germinate (grow into a seedling) immediately on reaching a suitable site. The period of inactivity prior to germination is called **dormancy**. Most seeds are produced towards the end of the summer and then have to undergo a period of dormancy (to allow for winter) before they will germinate. It is not always clear what triggers germination but in many cases a **period of time at a low temperature** followed by an increase in temperature is essential. The temperature change probably 'indicates' to the seed the onset of spring and favourable conditions.

Germination

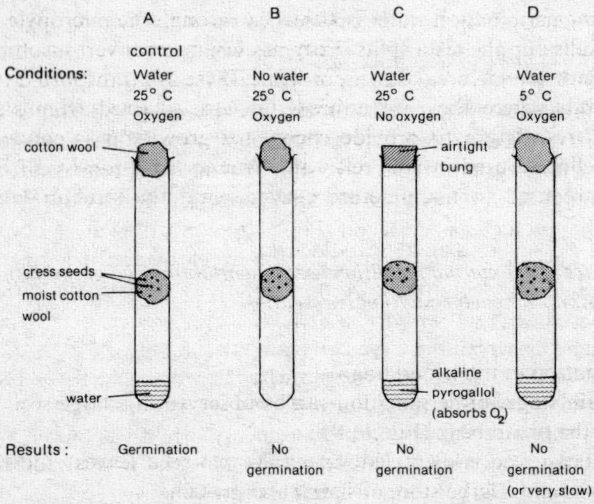

Fig. 11.7. *Experiment to investigate the conditions necessary for germination to occur*

Definition: Germination *is the growth of the seed into a seedling* (*a young plant able to photosynthesize*).

The conditions necessary for germination, following a period of dormancy, are:

1. water supply
2. suitable temperature
3. oxygen supply.

In addition, certain strains of grass seed require light before they will germinate. Most seeds, however, are indifferent to light intensity.

Before growth can occur insoluble food reserves, notably proteins and starch, must be broken down to soluble foods. Water is a reactant in these hydrolysis (breakdown) reactions and the enzymes that catalyse them require water and a suitable temperature. Oxygen is required for aerobic respiration to supply the energy for anabolic reactions.

The requirement for these conditions for germination to occur can be shown in a simple experiment using cress seeds (see Fig. 11.7). In this experiment, germination occurs only in the control seeds – those seeds provided with water, oxygen and a suitable temperature. If any of these factors are missing (as in tubes B, C and D), the seeds do not germinate. The conclusion is that oxygen, water and a suitable temperature are required for germination to occur.

During germination water is absorbed through the micropyle and the seed swells and the testa splits. Enzymes begin to convert insoluble food stores into soluble breakdown products. These are transported to growing points where they are utilized. Glucose (derived from starch) is used in respiration to provide energy for growth or is condensed to form cellulose used in new cell walls. Amino acids (derived from proteins) are used to manufacture enzymes and the structural parts of cells.

Q. What three *external conditions are essential for seed germination? Give* one *reason why each condition is essential.*

Germination in the broad bean

The main stages of germination can be observed in a large seed such as that of the broad bean (Fig. 11.8).

The bean seed is dicotyledonous (has two seed leaves) and the seed leaves contain a large store of starch and protein.

After the seed swells and the testa splits, the radicle (young root) grows from the region of the hypocotyl down into the soil by positive geotropism (page 186). The plumule (young shoot) grows from the epicotyl region upwards by negative geotropism. The young root develops root hairs (page 312) which absorb water and mineral salts from the soil. Later lateral roots grow off the radicle and provide anchorage for the seedling. When the plumule breaks through the ground

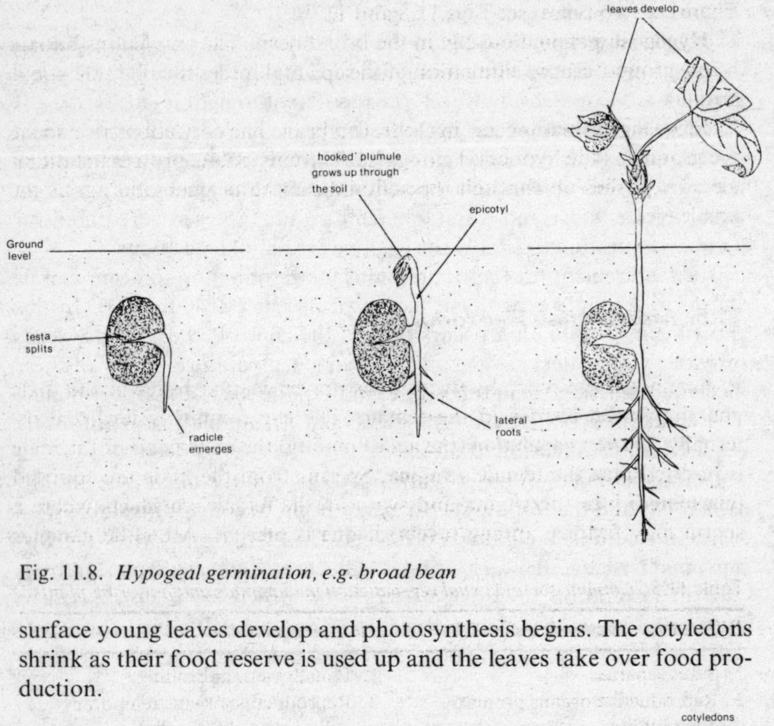

Fig. 11.8. *Hypogeal germination, e.g. broad bean*

surface young leaves develop and photosynthesis begins. The cotyledons shrink as their food reserve is used up and the leaves take over food production.

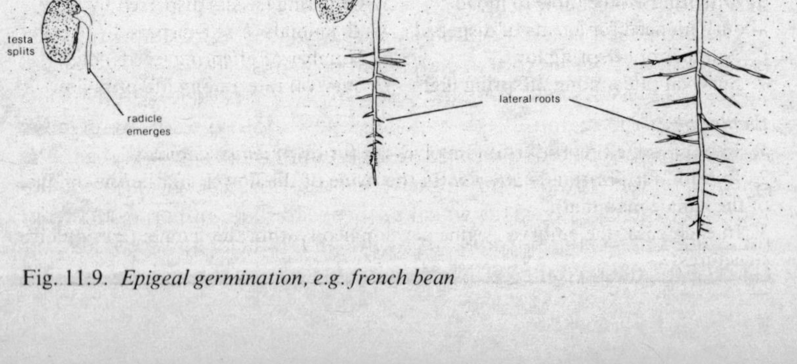

Fig. 11.9. *Epigeal germination, e.g. french bean*

Types of germination

There are two types (see Figs 11.8 and 11.9):

1. **Hypogeal germination**, e.g. in the broad bean. The cotyledons remain below ground. Rapid elongation of the epicotyl forces the plumule above ground.
2. **Epigeal germination**, e.g. in the french bean. The cotyledons rise above the ground as the hypocotyl elongates. The cotyledons protect the plumule as it pushes up through the soil and they turn green and act as the first leaves.

11.5. *Mammalian Reproduction*

In mammals, the reproductive organs are the paired **testes** in the male and the paired **ovaries** in the female. The egg (ovum) is fertilized internally. During **copulation** (the act of mating) the erect **penis** of the male is inserted into the female's **vagina**. Sperms from the male are squirted (**ejaculated**) into the vagina and swim up the female's **oviducts** where a sperm may fertilize an egg (**ovum**) if one is present. After the gametes

Table 11.5. *Comparison of sexual reproduction in mammals and flowering plants*

Differences: Mammals	*Flowering plants*
1. Sexes separate	1. Usually hermaphrodite
2. Reproductive organs present throughout life	2. Reproductive organs temporary (usually renewed annually)
3. Male gamete – active, transferred by mating, swims to female gamete	3. Male gamete – passive, transferred by insects, or wind. Male nucleus reaches egg nucleus via pollen tube
4. Continual development of embryo	4. Embryo often enters a dormant period after initial development
5. Offspring motile (able to move about); no need for agents of dispersal	5. Offspring (seeds) dispersed by wind, animals or self-dispersal
6. Number of offspring low	6. Number of offspring (seeds) high
7. Survival rate among offspring high	7. Survival rate among offspring low

Similarities:

1. In both, sexual reproduction involves the fusion of *unlike* gametes.
2. In both, fertilization is *internal* (in the ovule of the flower, and in the oviduct of the female mammal).
3. In each case, the embryo begins development within the female reproductive structures.

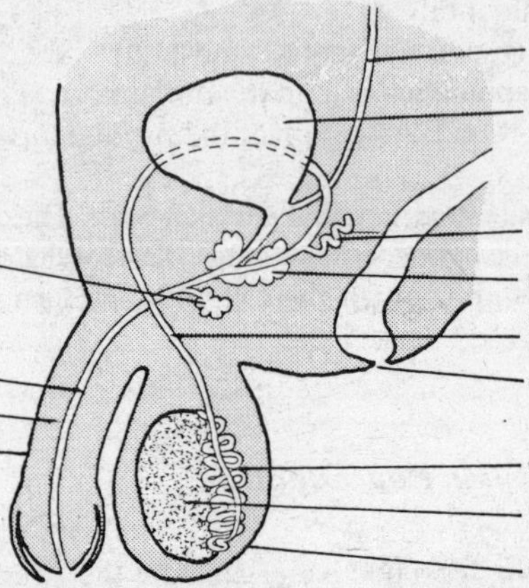

Fig. 11.10. *Human male reproductive organs* (*side or lateral view*)

fuse, the zygote formed undergoes mitotic cell division (page 251) and differentiates to become an embryo. Once organs are visible in the embryo it is called a **foetus**. The foetus remains in the female reproductive structures and when fully developed (after 40 weeks in humans), the female gives birth to the baby.

We will use human reproduction as our example: this follows the same overall plan as that in other mammals, but differs in the following points (compare with the rabbit, page 329):

1. The uterus of the human female is single, not branched.
2. Only one egg is normally shed at ovulation. Only one embryo develops at a time.
3. Development of the embryo takes a long time (40 weeks) in relation to the size of humans as mammals.
4. The reproductive cycle in women is called the menstrual cycle and lasts about 28 days. In most other mammals, if fertilization does not take place, the lining of the uterus is gradually reabsorbed. In the human female, the uterine lining breaks down and leaves the body as the **menstrual flow** (page 232).
5. Unlike other mammals, the human female does not have a period of 'heat' (highly receptive to males) at or around ovulation. Rather, the human female is potentially receptive at most times.

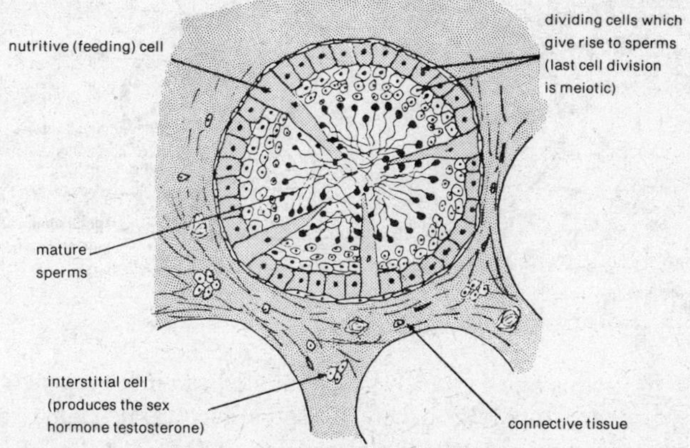

Fig. 11.11. *T.S. of a seminiferous tubule* (*highly magnified*)

Reproductive organs

The male reproductive organs, the **testes**, lie outside the abdominal cavity in a sac called the **scrotum** (see Fig. 11.10). Here they are at a slightly lower temperature than the rest of the body – this favours sperm production. Within the testes, sperms are produced in coiled tubes called **seminiferous tubules**. Each tubule is lined with actively dividing cells which

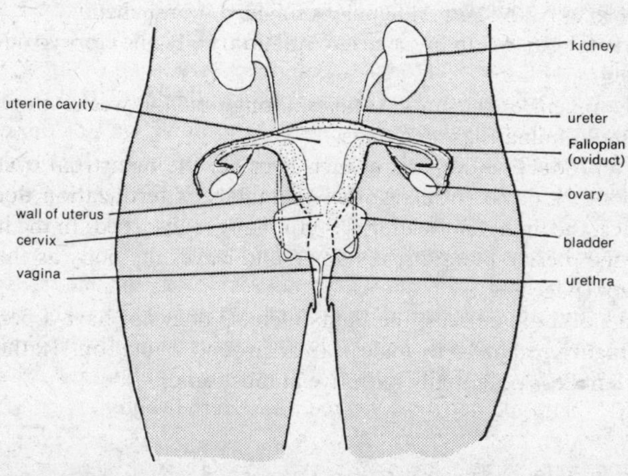

Fig. 11.12. *Human female reproductive organs* (*ventral view*)

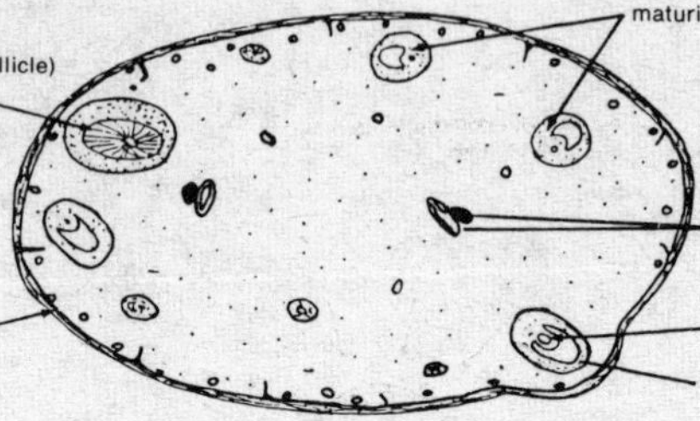

Fig. 11.13. *L.S. of ovary* (*slightly magnified*)

give rise to sperms (see Fig. 11.11). The tubules lead into the **epididymis**, where sperms are stored. From here, the **sperm duct (vas deferens)** leads to the **urethra** running through the **penis**. The urethra at different times carries urine (see page 155) and sperm.

Emptying into the vas deferens is the seminal vesicle which produces seminal fluid, the fluid in which sperms are suspended. The prostate gland produces secretions which activate the sperms to swim and provides them with nutrients. The penis contains many blood spaces which, during copulation (mating), become filled with blood at high pressure. This erects the penis, allowing it to penetrate the female's vagina during copulation.

The female reproductive organs, the **ovaries**, are two oval structures at the back of the abdomen (Fig. 11.12). They produce eggs (**ova**). From puberty onwards (see page 235) one or other of the ovaries releases an egg (ovum) about every 28 days. Inside the ovary, each ovum develops within a structure called the **Graafian follicle** (see Fig. 11.13). As the ovum matures the follicle moves towards the outer wall of the ovary. Finally, the follicle bursts, releasing the ovum, a process called **ovulation**. Once released, the ovum passes down the **oviduct (Fallopian tube)** propelled by the beating action of cilia in the tube lining. Subsequent events depend on whether the egg is fertilized or not.

The oviducts lead into the wide muscular tube, the **uterus (womb)** where, if the egg is fertilized, development of the embryo takes place. A ring of muscle, the **cervix**, separates the uterus from the **vagina**. The muscular vagina leads to the outside and, during copulation, receives the sperms. At birth, the baby is delivered through the vagina.

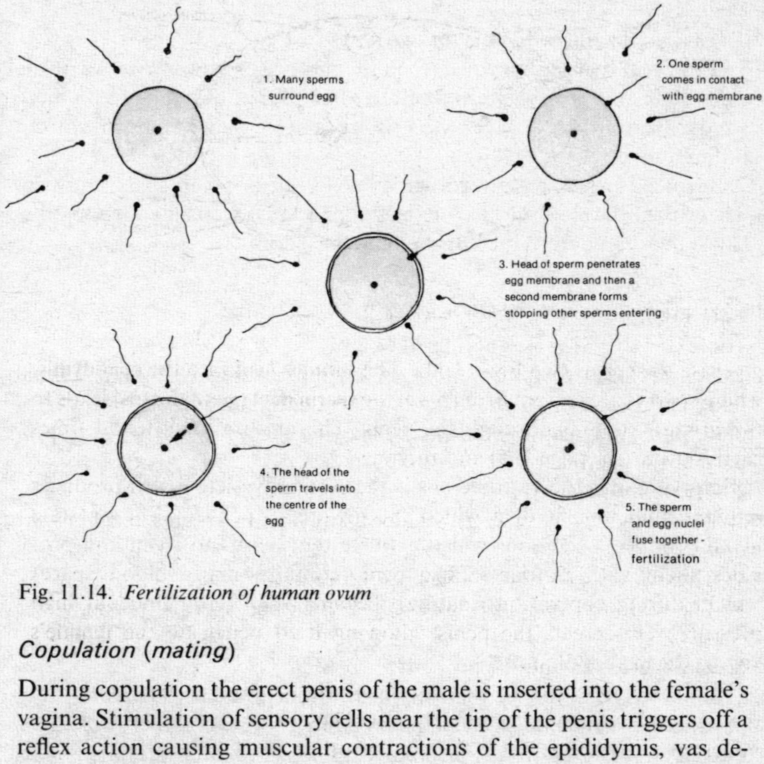

Fig. 11.14. *Fertilization of human ovum*

Copulation (mating)

During copulation the erect penis of the male is inserted into the female's vagina. Stimulation of sensory cells near the tip of the penis triggers off a reflex action causing muscular contractions of the epididymis, vas deferens and urethra. This results in **semen** (sperms suspended in seminal fluid and prostate gland secretion) being squirted (ejaculated) into the top of the vagina.

Fertilization

Active sperms swim through the cervix and up the oviducts, where they live for 2 or 3 days. A single ejaculation usually contains over 100 million sperms but few will enter the oviduct. If an ovum is present in one of the oviducts it may be fertilized by *one* of the sperms (see Fig. 11.14). As an unfertilized egg lives for only about 24 hours, there are only 2 or 3 days in the month when fertilization is possible. Fertilization occurs when the head of the sperm penetrates the egg membrane and the nucleus of egg and sperm fuse. A membrane develops around the fertilized egg to prevent other sperms entering (see Fig. 11.14).

Twins

Twins may be formed in one of two ways:

1. **Fraternal** twins are formed when the mother produces *two* ova at the same time and both are fertilized. The babies that develop are no more similar than normal brothers and sisters because they were formed from different ova.
2. **Identical** twins are formed when *one* fertilized ovum splits into two. These then develop into two babies which are genetically identical because they come from the same ovum and sperm.

Infertility

There are several causes of infertility:

1. Sperms need to be produced at a temperature lower than normal body temperature, which is why the testes hang outside the body. Tight underpants may cause infertility by keeping the testes too warm to produce fertile sperms.
2. The testes may produce sperms that are not formed properly.
3. The vas deferens may be blocked, preventing sperms reaching the urethra.
4. The ovaries may not be able to produce fertile ova.
5. The oviducts may be blocked, preventing sperms reaching the eggs.
6. The uterus may not be properly formed

Some women are given fertility drugs to help them to ovulate i.e. to produce ova. This may lead to the production of several ova at the same time and, if they are all fertilized, this leads to multiple births (e.g. quintuplets, sextuplets).

If successful fertilization cannot take place inside the woman it is sometimes possible to remove an ovum and fertilize it under laboratory conditions. The fertilized ovum is then placed inside the uterus to continue its development. If successful, this results in a so-called 'test tube' baby.

Gestation

The **gestation period** is the time from fertilization to birth and is around 40 weeks in humans. After fertilization the egg (now a **zygote**) divides many times by mitosis to become a hollow ball of cells. Now an **embryo**, it travels down to the uterus in about 4 to 7 days and then embeds itself in the uterine wall (**implantation**). Meanwhile, in the ovary, the Graafian

Table 11.6. *Development of embryo*

Time after fertilization	*Approx. size*	*Detail*
50 hours	Visible under microscope	Ball of cells
5 weeks	0.5 cm	Head beginning to develop. Simple heart. Limb buds visible.
10 weeks	3.5 cm	Placenta formed. Embryo now called a foetus. Most of its internal organs formed.
20 weeks	25 cm	Mother can feel movements of foetus. Heartbeat can be heard with stethoscope. Sex can be determined by ultrasound. Foetus is covered in fine hair called **lanugo**.
28 weeks	35–38 cm	Foetal head now usually above cervix. Finger and toe nails developed. Thin film of fat provides barrier between skin and amniotic fluid.
40 weeks	50 cm	Lanugo has disappeared. Ready for birth.

follicle which gave rise to the egg, has become an endocrine gland (see page 176) called the **corpus luteum**. This secretes the hormone, **progesterone**, which stops further egg production by the ovaries and stimulates an increase in blood supply to the uterus.

As the embryo begins to form, finger-like projections called **villi** grow from it into the uterus wall. These, together with part of the uterine wall, become the **placenta**, a special organ which supplies the embryo with food and oxygen. By 2 months, the embryo has developed limbs and main organ systems and it is now referred to as a **foetus**. It is connected to the placenta by the **umbilical cord**.

In the placenta the foetal blood circulation comes in close contact with the mother's (maternal) blood circulation. Thin membranes separate foetal capillaries from blood spaces in the uterine wall and some substances can diffuse freely across (see Fig. 11.15). Oxygen and dissolved foods diffuse from the mother's blood to the foetal blood. Carbon dioxide and soluble excretory products such as urea, produced by the foetus, diffuse in the other direction. The villi give the uterus and placenta a greater surface area for the exchange of substances. The placenta also acts as a selective barrier preventing certain harmful substances reaching the foetus. In addition, the placenta produces the hormones **oestrogen** and **progesterone** which prevent menstruation (page 232) and maintain the blood supply in the uterine wall.

The foetus is surrounded by a 'water sac', the **amnion**, which contains

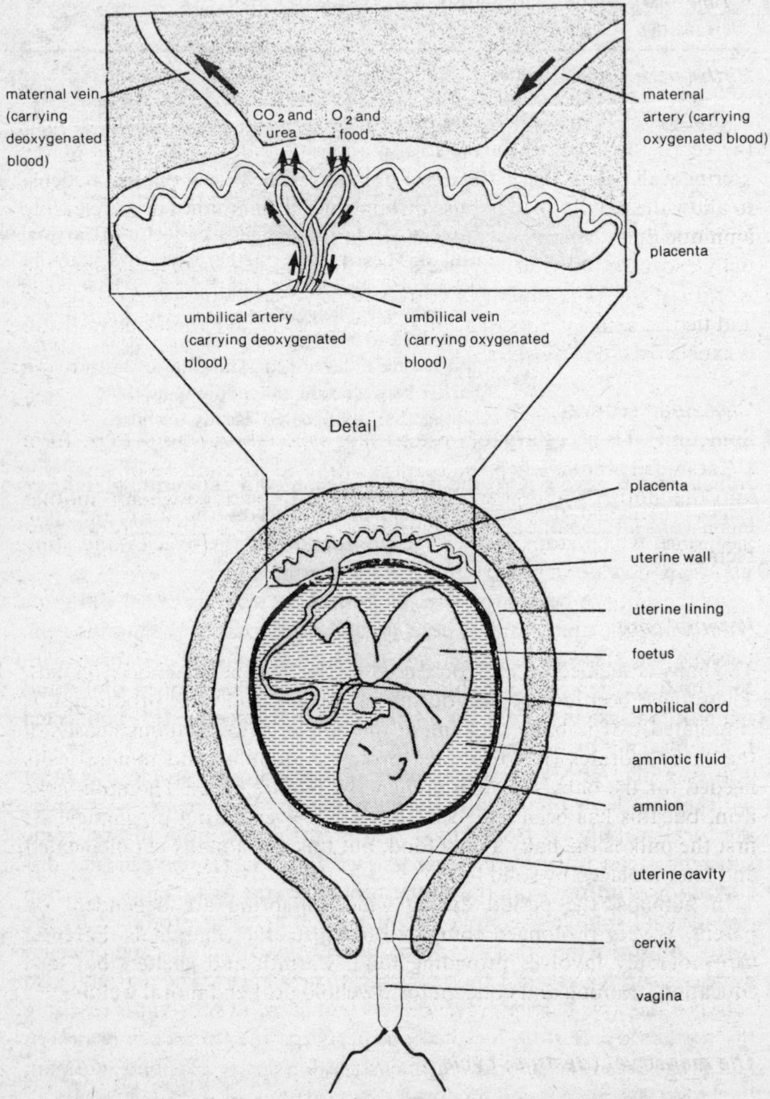

Fig. 11.15. *The foetus a few weeks before birth, with detail showing placenta*

amniotic fluid. This serves to protect the foetus from buffeting and sudden temperature changes, and allows the foetus to move.

Birth (parturition)

A few weeks before birth, the foetus comes to lie head down, towards the cervix. Just before birth, strong and rhythmic contractions of the uterine wall begin; this is the onset of labour. The cervix dilates (widens) to allow the baby's head to pass through and the amnion bursts releasing amniotic fluid. Muscular contractions of the uterus and abdomen eventually expel the baby, usually head first, through the vagina and into the world outside. The umbilical cord, still attached to the placenta, is cut and tied. A little later the placenta comes away from the uterine wall and is expelled as the 'afterbirth'.

Caesarean section

Sometimes it is necessary for obstetricians (specialist doctors) to perform a **Caesarean section** when the baby is extracted through an opening cut into the mother's abdomen. This may have to be done when a mother has a very large baby and a small pelvis, or when there are other problems.

Parental care

The baby is suckled on milk produced by the mother's **mammary glands**. These have been developing during pregnancy and milk production is stimulated by the baby suckling at the nipple. Milk contains nearly all the carbohydrates (lactose), fats, proteins, vitamins and mineral salts needed for the baby's growth and energy requirements. The milk lacks iron, but this has been stored in the baby's liver during pregnancy. At first the milk is the baby's only food, but this is gradually supplemented and later replaced by solid food.

In humans, the period during which offspring are dependent on parents is very prolonged compared to most other mammals. Parental care not only involves providing food, warmth and shelter, but also education, training and concern for psychological and moral welfare.

The menstrual (oestrus) cycle

NOTE. Details concerning how the cycle is controlled by hormones are not required by all examination boards. Check with your syllabus.

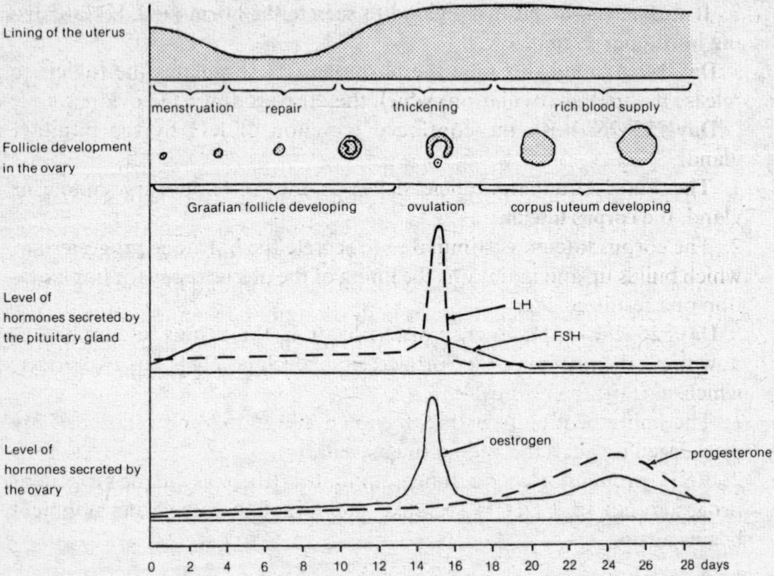

Fig. 11.16. *The menstrual cycle*

The reproductive cycle in women is known as the **menstrual** or **oestrus** cycle. Oestrus is the period of ovulation (when an egg is released). Menstruation is when the lining of the uterus is shed. The female reproductive cycle repeats itself about every 28 days from puberty (age 10–15) until menopause (at about age 50). The cycle is interrupted only by pregnancy.

The menstrual cycle, as shown in Fig. 11.16, is regulated by hormones secreted by the pituitary gland and ovary. During pregnancy, hormones secreted by the placenta delay recurrence of the cycle.

Normal cycle

Day 1: Menstruation begins and the pituitary gland (page 176) is stimulated to produce the hormone FSH (follicle stimulating hormone).

Days 5–14: FSH stimulates development of a Graafian follicle within the ovary. The follicle, in turn, secretes the hormone **oestrogen** which has a number of effects:

1. It causes the lining of the uterus to thicken (in preparation for implantation).
2. It inhibits the production of FSH by the pituitary gland (this, in turn, stops other follicles developing in the ovary).

3. It stimulates the pituitary gland to secrete the hormone LH (luteinizing hormone) – see below.

Day 14: Around this time the hormone LH stimulates the follicle to release its ripe egg (ovulation) which then travels down the oviduct.

Days 15–26: With the continued secretion of LH by the pituitary gland:

1. The burst Graafian follicle develops into a temporary endocrine gland, the **corpus luteum**.
2. The corpus luteum is stimulated to secrete the hormone **progesterone**, which builds up and maintains the lining of the uterus ready for implantation of a fertilized egg.

Days 26–28: In the absence of fertilization, the corpus luteum breaks down and the production of progesterone declines. This has two effects which start off the next cycle:

1. The lining of the uterus breaks down and blood and cell debris are discharged through the vagina (menstruation).
2. FSH production is no longer inhibited (there is no oestrogen or progesterone) so FSH is secreted and Graafian follicle development begins again.

Pregnancy cycle

If successful fertilization occurs:

1. The embryo implants in the uterus wall.
2. The corpus luteum persists, secreting progesterone, which maintains the uterus throughout early pregnancy.
3. The mammary glands enlarge.
4. Later in pregnancy, the placenta takes over the role of the ovary in secreting progesterone and oestrogen.

Just before birth, the level of progesterone falls, allowing oestrogen and the pituitary hormone, **oxytocin**, to take effect and cause contractions of the uterus which expel the baby through the cervix and the vagina.

After birth, a decrease in progesterone and oestrogen levels occurs (because the placenta is expelled) and this causes FSH to be released again by the pituitary gland. After some time, a return to the menstrual cycle follows.

Suckling causes another pituitary hormone, **prolactin**, to be released, and this causes secretion of milk from the mammary glands.

Notice how the reproductive cycle is governed by the pituitary gland and by feedback (see page 178) between four hormones, FSH, LH, oestrogen and progesterone.

Table 11.7. *Methods of birth control*

Method	*Description*	*Comments*
Condom (sheath, rubber, Durex, french letter)	Made of very thin rubber. Rolled over erect penis before intercourse. Sperms caught in end.	Must be put on before penis contacts vulva and removed carefully. More effective when used with spermicide.
Pill (oral contraceptive)	Tablets containing hormone mixture that prevents ovulation. Taken by woman every day for 21 days. She then stops for 7 days during which time she has a period.	Very effective. Only available on prescription. Not advised for women suffering from heart or liver disease or diabetes.
Intra-uterine device (I.U.D.) (coil or loop)	Piece of plastic or metal placed in uterus by doctor. Prevents implantation.	Very effective. Not suitable for all women. May cause heavier periods in some younger women.
Diaphragm (Dutch cap)	Dome-shaped disc of thin rubber placed at top of vagina before intercourse. Smeared with spermicide before fitting.	Very effective when fits cervix properly. Should not be removed before 6–8 hours after intercourse.
Spermicides	Chemicals in cream, jelly, foam or tablet form that kill sperms. Cream may be put in sheath or around diaphragm. All forms may be placed high in vagina not more than 10 minutes before intercourse.	Not very effective on their own. Best when used to improve safety of condom or diaphragm.
Sterilization	Woman–oviducts are cut to prevent ova reaching uterus. Man–vas deferens are cut to prevent sperms reaching urethra (vasectomy).	Extremely effective. Only suitable for couples who already have a family as is not usually reversible. Men may continue to produce live sperms for up to 6 months.
Rhythm method (mucothermic or calendar method)	Woman records body temperature daily and looks to see when thin, clear mucus is discharged from vagina. Mucus becomes clear just before ovulation and there is a slight temperature increase at the time of ovulation. The woman must avoid sex around the time of ovulation.	Not very effective because the temperature increase is very small and requires accurate measurements. Also the 'safe period' varies with the individual. Sperms and ova can stay fertile for varying lengths of time.
Withdrawal	The penis is removed from the vagina before ejaculation.	Not at all reliable as a small amount of semen leaves the penis as soon as it erects.

Puberty

Puberty refers to the time of life when physical and psychological changes occur which transform a child into an adult capable of sexual reproduction. Puberty is triggered by hormones released from the pituitary gland. It is marked by the maturation of the reproductive organs and the development of secondary sexual characteristics. The reproductive organs (**gonads**) start to release the sex hormones, **oestrogen** (in females) and **testosterone** (in males). Secondary sexual characteristics refer to those physical features associated with sex (male or female) but not directly concerned with reproduction.

At puberty, the following physical changes occur:

In **girls** (age 10–15):

1. Ovaries mature and secrete oestrogen.
2. Menstruation starts.
3. Oestrogen stimulates development of the following secondary sexual characteristics:

(a) breast development
(b) fat deposits laid down under skin (giving the characteristic adult female body shape)
(c) hair grows under arms and in pubic region.

In **boys** (age 11–16):

1. Penis and testes enlarge, testes secrete testosterone.
2. 'Wet dreams' (nocturnal emissions of semen) begin.
3. Testosterone stimulates development of the following secondary sexual characteristics.

(a) body becomes more muscular
(b) larynx enlarges and voice deepens
(c) beard growth
(d) hair grows under arms and in pubic region.

11.6. *Family Planning*

There are many methods of birth control available, most commonly involving the use of **contraceptives**. These are devices or drugs that prevent conception.

The various methods of birth control are described in Table 11.7.

Table 11.8. *Sexually transmitted diseases*

Disease	*Cause*	*Early symptoms*	*Late symptoms*	*Treatment*
Syphilis (pox)	*spirochaete* bacteria	Sores on or inside sex organs.	Heart disease, insanity, paralysis and death.	Antibiotics, cure may take 2 years.
Gonorrhoea (clap)	*gonoccoccus* bacteria	Men – pus from penis, burning pain on passing water. Women – extra mucus from vagina. 70% have no symptoms.	Scars form inside sex organs, tubes may become blocked causing sterility.	Antibiotics
Vaginitis	*protozoa Trichomonas vaginalis*	Women only. Sore, itchy vulva and vagina, yellow, smelly discharge.	Early symptoms continue.	Chemicals to kill protozoa.
Pelvic inflammatory disease (P.I.D.)	bacteria	Women only. Burning pain in abdomen.	Oviducts may become blocked by scar tissue.	Antibiotics
Non-specific urethritis (N.S.U.)	not known	Men only. Pus from penis, pain when urinating.	Arthritis. Infections of skin, eyes and mouth.	Antibiotics. May recur.
Thrush	fungus	Rare in men, penis red and sore at tip. Women – sore, itchy vagina. Thick discharge. Not always transmitted sexually.	Discharge may worsen, other symptoms continue.	Fungicide applied as pessaries, ointment or tablets.
Herpes	virus	Sores like 'cold sores' on penis or vagina.	Possibly causes other damage in body. Can pass to baby in uterus.	Different methods, very difficult to cure.
A.I.D.S. (acquired immune deficiency syndrome)	virus	Frequent infections and illness due to suppression of immune system.	Body worn down by illnesses, eventually death.	None
Crab lice (crabs)	lice	Itchiness. Lice visible on pubic hair.	Lice spread to rest of body. Blood loss can weaken body, leading to secondary infection	Washing in special shampoo.

11.7. *Sexually Transmitted Diseases*

There are a number of diseases that can be passed from one person to another during sexual intercourse. These are known as **venereal diseases** (V.D.) and are described in Table 11.8.

Definitions

Reproduction
Haploid
Diploid
Fertilization
Hermaphrodite
Pollination
Germination

Key Words

Asexual reproduction
Sexual reproduction
Binary fission
Spore formation
Budding
Vegetative propagation
Runner
Sucker
Stem tuber
Bulb
Rhizome
Artificial propagation
Cuttings
Grafts
Scion
Stock
Self-fertilization
False fruit
Wind dispersal
Animal dispersal
Internal fertilization
External fertilization
Dormant
Zygospore
Gametophyte
Sporophyte
Receptacle
Calyx
Sepals
Corolla
Petals
Nectary
Stamens
Filament
Anther
Pollen grains
Carpels
Ovary
Sperm duct (vas deferens)
Urethra
Style
Stigma
Ovule
Embryo sac
Egg nucleus
Self-pollination
Protandry
Protogyny
Insect-pollinated
Wind-pollinated
Micropyle
Fruit
Seed
Radicle
Plumule
Cotyledon
Testa
True fruit
Placenta
Umbilical cord
Oestrogen

Self-dispersal
Seed dormancy
Hypogeal
Epigeal
Testes
Ovaries
Copulation
Penis
Ejaculate
Egg (ovum)
Foetus
Menstrual flow
Seminiferous tubules
Epididymis
Graafian follicle
Ovulation
Oviduct (Fallopian tube)
Uterus (womb)
Cervix
Vagina
Semen
Gestation
Embryo
Implantation
Corpus luteum
Progesterone
Villi
Amnion
Amniotic fluid
Mammary glands
Menstrual (oestrus) cycle
Oestrus
Menstruation
FSH
LH
Oxytocin
Prolactin
Puberty
Testosterone

Questions

1. (a) Describe sexual reproduction in a named *insect-pollinated flower referring in your answer to*
(i) pollination
(ii) fertilization, and
(iii) seed and fruit formation.
(b) In what ways does a wind-pollinated flower differ from your named example?

2. The table below shows the changes in weight of a developing baby

Weeks after fertilization	*5*	*10*	*15*	*20*	*25*	*30*	*35*	*40*
Weight (kg)	*0.1*	*0.15*	*0.2*	*0.45*	*0.75*	*1.5*	*2.5*	*3.5*

(a) Use these data to draw a graph showing the baby's development.
(b) How many weeks after fertilization does the baby weigh 2.0 kg?

3. Items A–C are contraceptives and (i)–(iii) are descriptions of contraceptives. Match the contraceptive with its description.

A. condom — *(i) prevents implantation*
B. pill — *(ii) prevents ovulation*
C. I.U.D. — *(iii) prevents sperms entering vagina.*

12. *Growth of Organisms and Populations*

A characteristic feature of living organisms is that they grow.

Definition: Growth *is an increase in the amount of protoplasm in an organism.*

The term protoplasm is used to distinguish between true growth and a simple increase in size due, perhaps, to an organism simply absorbing a lot of water.

Growth occurs when anabolic processes exceed catabolic processes (see page 20). Growth requires raw material and energy (from respiration).

The term 'growth' can be applied to an individual or to a population of individuals, in which case growth is increase in numbers.

In multicellular organisms, growth is accompanied by cell division, an increase in cell numbers, and an increase in the organism's complexity.

12.1. ***Growth Curves***

Populations of organisms, and individual organisms themselves, show a pattern of growth based on, or a slight modification of, the standard

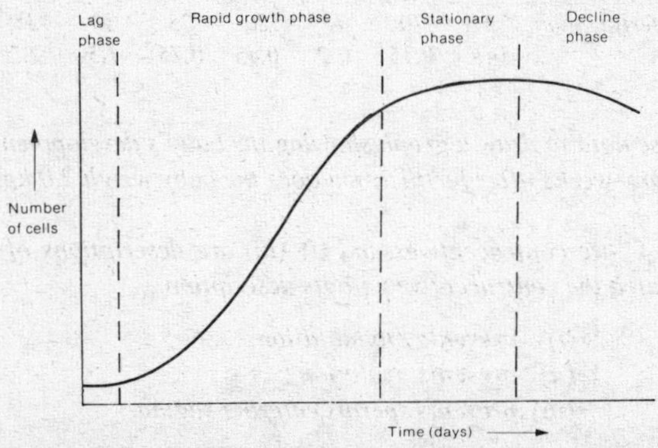

Fig. 12.1. *Standard growth curve for a microbial culture*

growth curve (Fig. 12.1). This curve describes well the growth and decline in numbers normally found in a culture of micro-organisms such as bacteria, protozoa, yeast or single-celled algae. During the **rapid growth phase** the population is increasing at a maximum rate for the species under the existing conditions. As they grow, the cells use up nutrients and produce waste products which accumulate in the medium. Eventually growth slows down and stops due either to a lack of nutrients or the build up of waste products (or both). This is the **stationary phase**. It leads into the **decline phase** in which the number of living cells in the population gradually decreases.

The standard growth curve is variously modified in multicellular organisms. The growth curve of an annual flowering plant (page 308) follows closely the standard form (see Fig. 12.2). In a perennial tree, e.g. the oak tree (page 309), growth stops in winter and resumes in spring

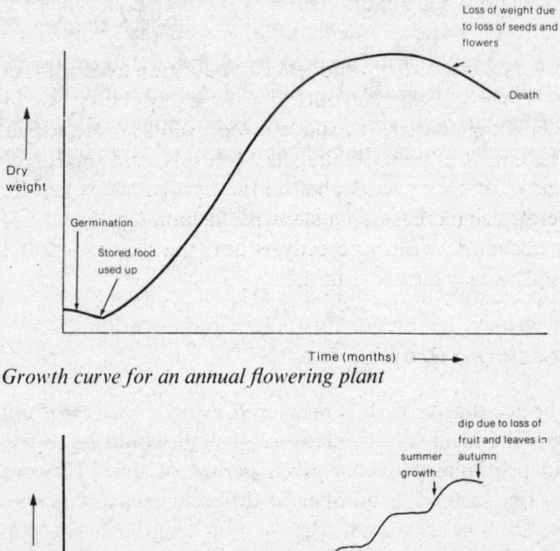

Fig. 12.2. *Growth curve for an annual flowering plant*

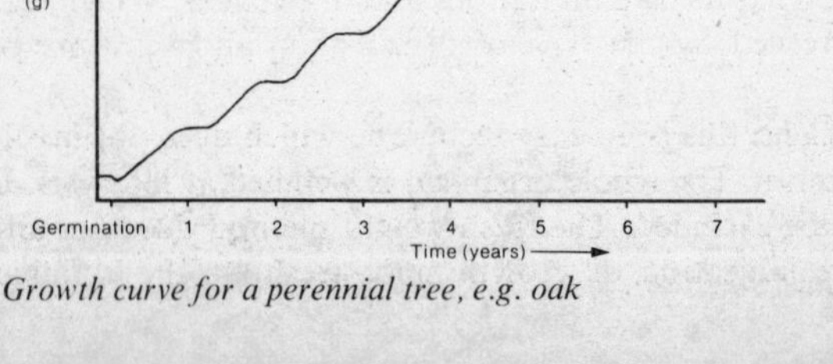

Fig. 12.3. *Growth curve for a perennial tree, e.g. oak*

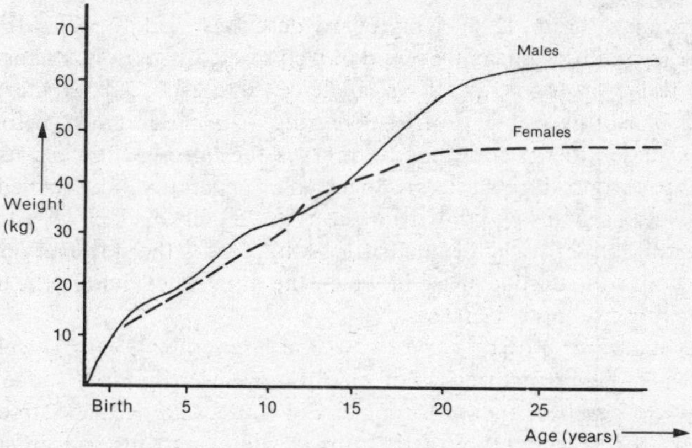

Fig. 12.4. *Human growth curves*

(see Fig. 12.3). Loss of fruit and leaves in autumn accounts for a loss in weight. In humans, there is a spurt of growth at puberty (see Fig. 12.4).

In insects, a rigid cuticle (exoskeleton) prevents any continual increase in size. The cuticle is periodically shed (**ecdysis**) and the insect swells in size (by intake of air or water) before the new cuticle is laid down. As a result, overall size increases in a stepwise fashion (see Fig. 12.5). Growth of tissues occurs between one ecdysis and the next, so that the insect 'grows to fit' its new cuticle (Fig. 12.5).

12.2. ***Measuring Growth***

You will notice that growth is measured using a variety of units of size or number. The ideal way to measure growth would be to measure the increase in protoplasm over a given period of time. However, this is difficult to do. Instead, a number of different measurements are used: increase in fresh (wet) weight, dry weight. length, or for populations, numbers. In all cases, the increase is measured over a set period of time. Each method has its own limitations and drawbacks. Which technique is used is governed by the type of organism(s) and the time period involved.

1. **Fresh weight:** this is an easy technique which does not involve injury to the organism. The whole organism is weighed at intervals. In plants, soil should be excluded. The fresh weight method does not always give an accurate indication of growth since fresh weight is influenced by

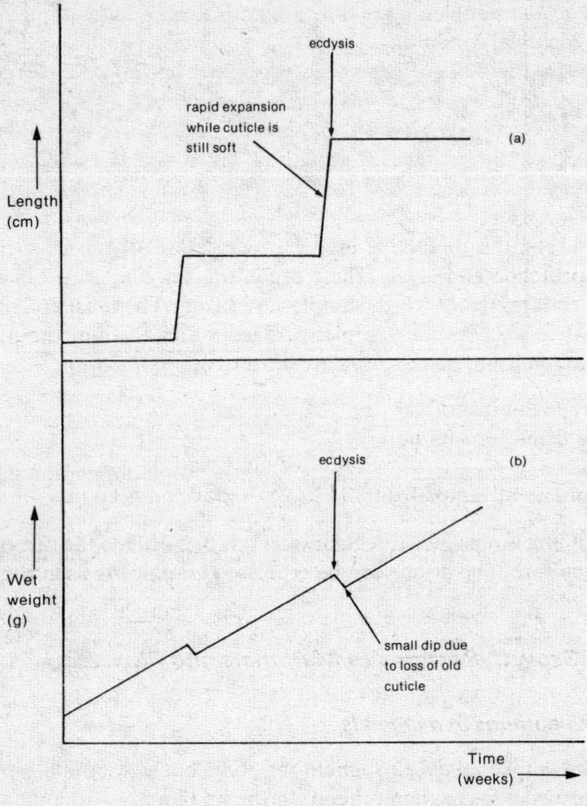

Fig. 12.5. *Growth curves for an insect showing incomplete metamorphosis, e.g. locust*

changes in water content of the body. It is used to measure growth in vertebrates.

2. **Dry weight:** this method usually gives a more meaningful result. It involves killing the sample by gently heating it to a constant weight to remove water. It is frequently used for seedlings. However, an increase in dry weight could be due to an increase in non-living material, e.g. lignin in plants. Also, it is possible for organisms to lose dry weight while still growing, as in the case of germinating seedlings which use up their food stores (see Fig. 12.2).

Valid results can be obtained using the dry weight method by taking

large *random* samples at set time intervals from a large population kept under identical conditions.
3. **Length:** this straightforward method does not injure the organism. It is most suitable for organisms whose growth is channelled mainly in one direction, e.g. the young root or shoot tips of flowering plants. Its disadvantage is that it ignores growth in other directions, such as width. This may be considerable, for example, the increase in girth of tree trunks.
4. **Numbers:** this method is used for unicellular organisms such as bacteria, protozoa and algae. These organisms may be grown in solutions called **cultures**, from which samples are taken. The most common technique is to take several samples of fixed volume, count the number in each sample, and then extrapolate for the total population:

$$\frac{\text{average number of organisms per sample}}{\text{volume of sample}} \times \text{total volume of culture} = \text{total population number}$$

Sampling is repeated at regular intervals to calculate the rate of growth of the culture. The results can be expressed graphically as in Fig. 12.1.

12.3. *Growth Patterns in Mammals and Flowering Plants*

Growth patterns in mammals

1. Growth takes place throughout the body, but may be **allometric** (some parts growing faster than others). In the human, for example, the brain grows slowly and stops growing at about age five. The legs, however, grow faster and keep growing until maturity.
2. Growth involves cell division and an increase in cell number, but not usually an increase in cell size.
3. Growth is **limited**, i.e. growth stops after a certain length of time. Growth processes may still occur, e.g. the production of new blood cells, but there is no overall increase in size.
4. Growth is, within limits, independent of temperature.

Growth patterns in flowering plants

1. Growth is usually restricted to regions called **meristems**. **Apical meristems** occur at root and shoot tips (see Fig. 9.12, page 184). The **cambium**

is a circle of actively dividing cells in stems and roots which is responsible for their increase in girth (see Fig. 15.15, page 309).
2. Cell division is followed immediately by rapid cell elongation due to the rapid uptake of water (Fig. 9.13, page 185). This is followed by **cell differentiation** (the cells becoming specialized for particular functions).
3. Growth is **unlimited**, i.e. a plant may retain the ability to increase its size throughout its life.
4. Environmental factors, e.g. seasonal temperature variations, influence growth rate.

Now try the following question:

Q. Briefly compare growth in a mammal with that in a flowering plant.

Similarities in growth between mammals and flowering plants

1. Growth is under the control of chemical substances (hormones).
2. The rate and extent of growth depends, in some measure, on the balance and quantity of food. In plants this will be affected by the supply of minerals, water and light, and on the temperature.
3. The growth curve for a mammal may resemble the same general form as that for a flowering plant (e.g. Fig. 12.2).
4. Growth is determined genetically (the genes influence growth through the action of enzymes and hormones).

12.4. *Metamorphosis*

Definition: Metamorphosis *is a major change in body form between the egg and adult stages of a life cycle.*

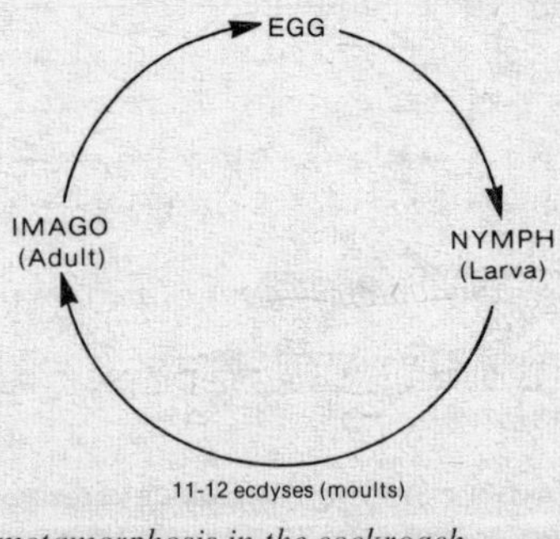

Fig. 12.6. *Incomplete metamorphosis in the cockroach*

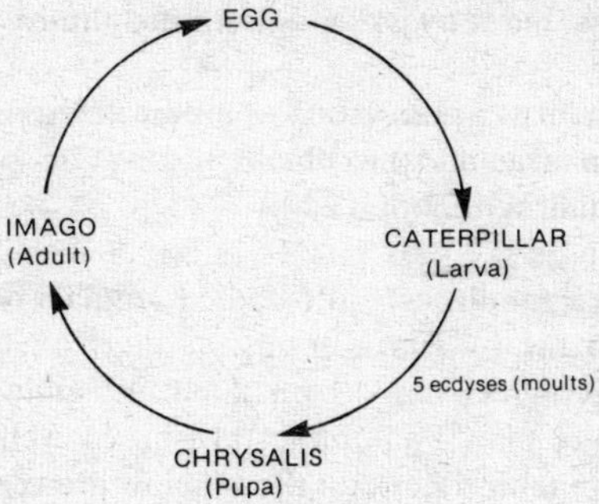

Fig. 12.7. *Complete metamorphosis in the butterfly*

The intermediate forms between egg and adult are called **larvae**.

Metamorphosis is found in insects (page 319) and amphibians (page 325). Insects fall into two groups based on their type of metamorphosis. Insects such as locusts and cockroaches show **incomplete metamorphosis** (rather misleadingly named). This means the larval form gradually changes into the adult over a series of moults when the cuticle is shed (Fig. 12.6). At the final moult the reproductive organs are mature and the wings are fully formed. The larval stages are called **nymphs**. Insects such as butterflies, moths and true flies show **complete metamorphosis**. A larval form strikingly different from the adult hatches out from the egg. The larva feeds actively and grows rapidly, shedding its cuticle several times, before changing into a **pupa** or **chrysalis**. It is during this apparently inactive stage that metamorphosis occurs; there is an extensive breakdown of larval tissues and reorganization into adult tissues. Eventually the adult or **imago** emerges from the chrysalis (see Fig. 12.7 and page 320).

In the butterfly, the larva is the major feeding phase of the life cycle, whereas the adult is the reproductive and dispersal phase. The two forms are specialized for different modes of life. The advantage of this is that there is no direct competition between larva and adult for space or food.

In amphibia, metamorphosis occurs when tadpoles develop into frogs or toads. Amphibians are only partially adapted for life on land and must return to water to breed. Metamorphosis is the transition between the aquatic larval phase, the tadpole, and the adult terrestrial phase, the frog or toad (see page 325).

12.5. *Population Growth*

The increase in growth of a population of multicellular organisms follows a pattern similar to that of a microbial culture (Fig. 12.1) but tends to level off rather than decline (Fig. 12.8).

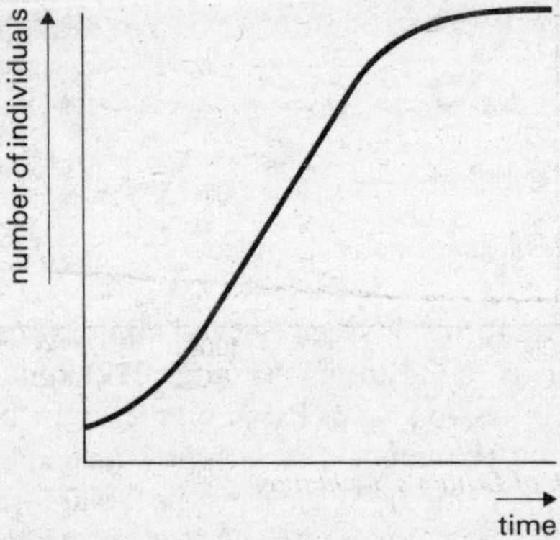

Fig. 12.8. *Population growth*

The population grows slowly at first but as the number of reproducing individuals increases the population grows faster.

Consider what happens if two rats (1 male, 1 female) are introduced to a barn. The breeding population is 2 to begin with but if they have a litter of 6 (3 males and 3 females), within a few weeks the youngsters will also be able to breed. The breeding population is now 8. If the 4 female rats each have a litter of 6 we now have a population of 32. These rats will soon breed again, leading to a population of 128. This type of increase in population is called **exponential**.

In reality, of course, population growth is not that simple. Some of the older rats will die, some of the young rats will not survive to adulthood. Eventually the population will level off due to **limiting factors**. The commonest limiting factors are listed below:

1. Availability of food.
2. Lack of space for nests.
3. Injuries and death caused by fighting due to overcrowding.
4. Infertility caused by the stress of overcrowding.
5. Disease, which spreads faster in crowded conditions.
6. Predation by other animals.

All populations are affected by such limiting factors to some extent.

Growth of the human population

Figure 12.9 shows how the population of Britain has grown.

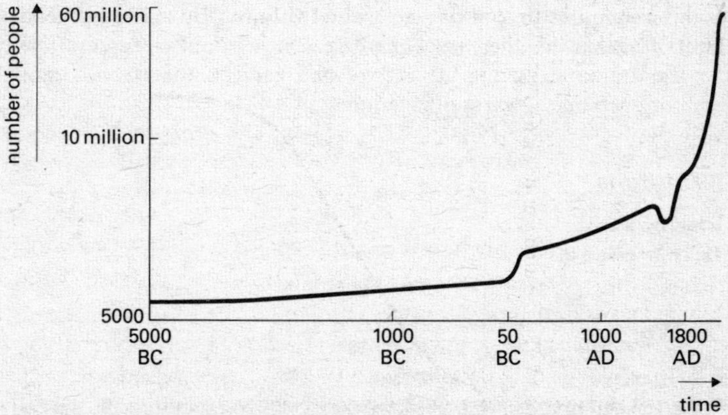

Fig. 12.9. *Growth of Britain's population*

The early growth of the population was very slow. At this time humans were hunter-gatherers, limited by the amount of food occurring naturally.

The development of agriculture allowed humans more control over their food supply and the population was able to grow faster. Even then the population was still subject to limiting factors, such as the Black Death of the Middle Ages.

Towards the end of the eighteenth century a priest called Thomas Malthus became concerned that the population was growing faster than the increase in food supply. He predicted mass starvation in Britain if the population continued to increase at the same rate. In fact, advances in agricultural technology were able to increase the food supply so dramatically that this did not happen.

Since the late nineteenth century the population of Britain has increased dramatically. This is due to a number of factors:

1. Health care has improved; more babies survive and people live longer.
2. The standards of housing and sanitation have improved, so there is less disease.
3. Improved diet has led to better health.
4. Food production is more efficient due to:

(a) the use of pesticides to control the insects that eat our crops

(b) improved harvesting methods

(c) improved methods of food storage
(d) new strains of crops that produce a higher yield.

The population of Britain is now beginning to level off, mainly because of the use of **birth control** (see section 11.6). In many developing countries, however, this is not the case. Their populations continue to increase dramatically and in years of bad harvests there is not enough food for everyone. The result is famine.

Definitions

Growth
Metamorphosis

Key Words

Growth curve
Rapid growth phase
Stationary phase
Decline phase
Ecdysis (moult)
Fresh (wet) weight
Dry weight
Exponential
Allometric
Limited growth
Meristem
Apical meristem
Cambium
Cell differentiation
Unlimited growth
Incomplete metamorphosis
Nymph
Complete metamorphosis
Pupa (chrysalis)
Imago

Questions

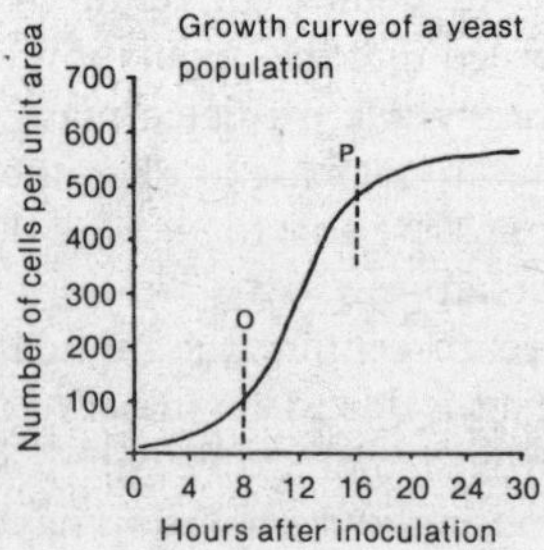

1. The curve on the graph above is typical of a population growth curve found in many organisms other than man.
(*a*) *State* one *fact about the yeast population up to point O on the graph that could explain the curve to that point.*

(*b*) *State* one *fact about the yeast population between points O and P on the graph to explain the shape of the curve between these points.*
(*c*) *List* three *factors which might cause the levelling-off as seen 15–24 hours after inoculation.*
2. The graph shows the average body mass of boys and girls up to the age of eighteen years.

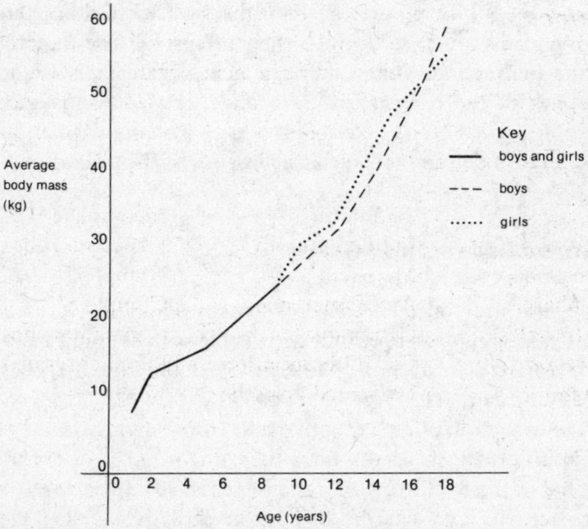

(*a*) *What is the average mass of six-year-old boys?*
(*b*) *Which have the greater average body mass (boys or girls)*
(*i*) *at age twelve?*
(*ii*) *at age eighteen?*
(*c*) *Suggest a reason for your answer to* (*b*).
(*d*) *What is the percentage increase in body mass of boys between age nine and age ten?*
3. Refer to the graph showing the growth of Britain's population (Fig. 12.9).
(*a*) *Why was the growth of the human population slow between 5000 and 1000 B.C.?*
(*b*) *What was the cause of the sudden decline of the population in the seventeenth century?*
(*c*) *What changes allowed the human population to grow so rapidly during the last century?*
(*d*) *Why is Britain's population now beginning to level off?*

13. Genetics and Evolution

Genetics is the study of the mechanisms by which characteristics of parent organisms are passed on to their offspring. The material which carries the instructions for building a new organism is found in the **chromosomes** of the cell nucleus (see page 21). The substance within chromosomes called DNA determines which proteins are manufactured, and these in turn determine the characteristics of the organism.

13.1. *Chromosomes and Genes*

Chromosomes are coiled strands of protein and DNA present in the nucleus. Under a microscope, they are visible with staining methods just before or during cell division. At other times they are dispersed in the nucleus as chromatin (see Fig. 3.3, page 24).

With the exception of reproductive cells (those giving rise to gametes), all cells in an organism's body have the same number of chromosomes (in man, 46; in the fruit fly, 8; and in rye grass, 20). In each of these cells there are actually two sets of chromosomes. This is called the **diploid** number of chromosomes (represented by 2n): in man 2n = 46.

Gametes contain half the diploid number of chromosomes: in humans, sperm and eggs have 23 chromosomes. This is called the **haploid** number and is represented by n.

Chromosomes are normally found in pairs which have a similar appearance. These are called **homologous pairs**. For example in the body cells of man there are 23 homologous pairs. In sexually reproducing organisms one member of each pair is derived from one parent, and the other member from the other parent.

Chromosomes have two main roles:

1. They carry instructions for controlling day-to-day activities within the cell.
2. They carry instructions for producing a new cell (or in the case of a gamete, half the instructions for producing a new organism).

Each chromosome has a large number of **genes** running along its length. A gene is a short section of a chromosome (or more precisely, a short section of DNA). Each gene carries a coded message for manufacturing a specific protein. Genes control characteristics of the organism by providing instructions for manufacturing particular proteins. In a body cell, there are two copies of each gene (one on each chromosome of a homologous pair) so that a given characteristic, e.g. eye colour, is controlled by at least one pair of genes.

13.2. *Mitosis and Meiosis*

During cell division it is important that an accurate copy of the chromosomes, and the genes they carry, is passed on to daughter cells.

Mitosis is the type of cell division associated with growth and asexual reproduction. It occurs in somatic (non-reproductive) cells. A parent cell divides into two daughter cells which have identical chromosomes. The parental cell is diploid (2n) and the resulting daughter cells are diploid (2n). During the division process the number of chromosomes is doubled by **replication**. The total chromosome material then divides equally between the two daughter cells making them identical (see Fig. 13.1).

Mitosis is a continuous process, but it can be broken down into a number of stages. For some examination boards you are required to know the different stages of mitosis. Check with your syllabus.

Meiosis is the type of cell division associated with sexual reproduction. It occurs in the cells which give rise to gametes. Meiosis has two important functions:

1. It produces daughter cells (gametes) which have half the chromosome number of the parent cell.

We have seen earlier (page 208) why this is so important for a sexually reproducing species. It is essential that gametes have half the chromosome number of a normal cell, so that when two gametes fuse to form a zygote, the normal (diploid) chromosome number is restored.

2. Meiosis produces daughter cells which have a combination of chromosomes different to that of the parent cell and to each other. This results in gametes which are genetically different and helps to produce genetic variation among offspring.

Meiosis generates variation by two processes:

(a) crossing-over

(b) the random assortment of chromosomes.

Meiosis involves two cell divisions, one after the other. This results in

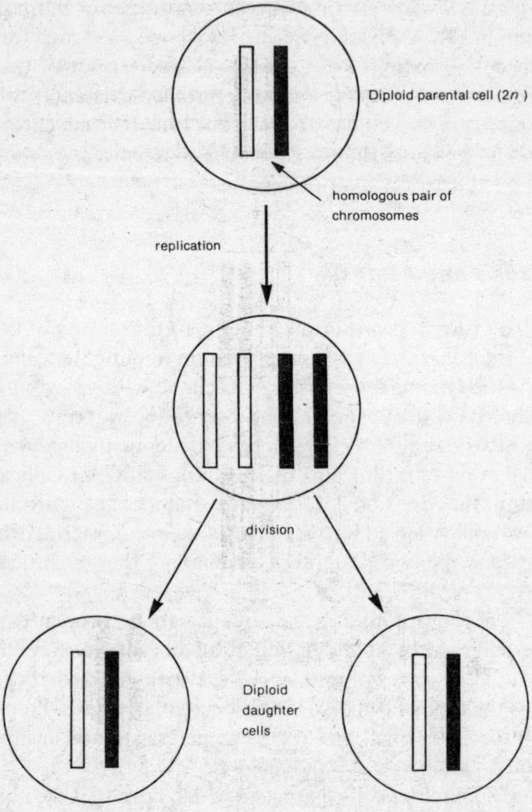

Fig. 13.1. *Mitosis (only one pair of homologous chromosomes shown)*

one parental cell giving rise to four daughter cells. Refer to Fig. 13.2 as you read through the text.

Before the first cell division, as in mitosis, the number of chromosomes is doubled by **replication** (duplication of chromosomes). In meiosis, however, homologous pairs of chromosomes come to lie side by side and attach at certain points called **chiasmata**. At these points genetic material can be exchanged between chromosomes by a process called **crossing-over**. The result is that a particular chromosome may come to contain some sections (and hence genes) from the male parent, and some from the female parent.

During the first division of meiosis, homologous pairs of chromosomes

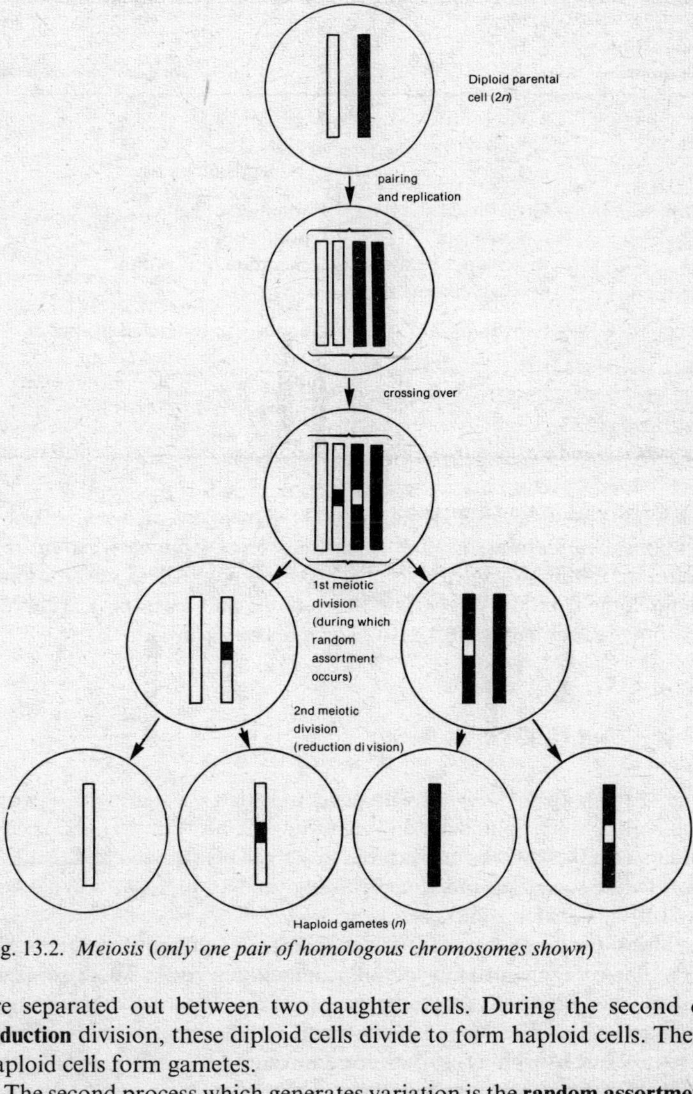

Fig. 13.2. *Meiosis* (*only one pair of homologous chromosomes shown*)

are separated out between two daughter cells. During the second or **reduction** division, these diploid cells divide to form haploid cells. These haploid cells form gametes.

The second process which generates variation is the **random assortment of chromosomes**. During the first division of meiosis, which chromosome of a homologous pair goes to which daughter cell is determined at

Table 13.1. *Summary comparison of mitosis and meiosis*

	Mitosis	*Meiosis*
Number of cell divisions	1	2
Resulting cells	2 cells formed diploid (2n) identical	4 cells formed haploid (1n) non-identical
Purpose	Growth and replacement. In certain organisms asexual reproduction	Formation of gametes in sexual reproduction. Generation of variation
Occurrence	Growth regions	Gonads, e.g. testes and ovaries of mammals (see pages 224 and 225); anthers and ovules of flowering plants (see page 214). Sporangia of ferns (see page 305)

random. The result is that at the end of meiosis, each cell (gamete) contains a random assortment of chromosomes from each parent. For example, a human gamete has 23 chromosomes. All 23 chromosomes might come from one parent, or 10 from one and 13 from the other, or any other combination of the two which makes up 23.

13.3. *What the Words Mean*

One of the biggest problems with genetics is that there are lots of words that sound similar but mean different things. Unfortunately you need to understand these words and be able to use them correctly. It is therefore worthwhile spending some time learning the terminology to avoid confusion. This section should help you.

Alleles. Each gene for a particular character is present twice in diploid cells, one on each chromosome of a homologous pair. These genes are not always identical; e.g. some people are able to roll their tongue, others are not. The same gene is involved but there are two different versions, one for rolling and one for non-rolling.

These different versions are **alleles** of that gene.

Definition: Alleles *are different types of one gene. They affect the same character and are at the same site* (*locus*) *on the chromosome.*

Symbols for different alleles. Each gene can be represented by a letter, using capital and small forms of the letter for different alleles; e.g. We can use T for the allele for tongue rolling and t for the non-rolling allele.

Each person has two alleles of that gene so the possible combinations are TT, Tt and tt. When one of each is present, i.e. Tt, one allele may be **dominant** to the other. In this example T is dominant to t. Therefore t is said to be **recessive** to T.

Remember the capital letter must represent the dominant allele. So a Tt person can roll his or her tongue because there is a T present.

Heterozygous. This describes a person carrying different alleles of the gene, i.e. Tt. A person with Tt is a **heterozygote**.

Homozygous. This describes a person carrying the same alleles of a gene, i.e. TT or tt. Both of these are **homozygotes**.

Phenotype. A person who can roll his or her tongue may be TT or Tt. The ability to roll the tongue is the **phenotype**, so a tongue roller carries a T but it is impossible to tell, just by looking, if a t is also present.

Definition: Phenotype *is the character of the organism that can be observed.*

Genotype. Two tongue rollers TT and Tt have the same phenotype but their genetic make-up is different; that is they have different **genotypes**.

Definition: Genotype *refers to the genetic make-up of the individual.*

13.4. *Breeding Experiments*

In the 1850s a monk called Gregor Mendel carried out a number of breeding experiments to investigate the laws of inheritance. He worked with the sweet pea plant which has a number of varieties.

Two such varieties are produced by different alleles of the gene which controls the height of the plant. One allele produces tall plants and the other produces dwarf plants.

Mendel first established that the plants he was using were **pure breeding**. This means that when the plants were self-pollinated (**selfed**) the offspring showed the same characteristics. For example, pure breeding tall plants always produce tall offspring; pure breeding dwarf plants always produce dwarf offspring.

Mendel crossed a pure breeding tall plant with a pure breeding dwarf plant. All of the resulting seeds grew into tall plants. This first generation is known as the F_1 (first filial) generation. When the F_2 plants were

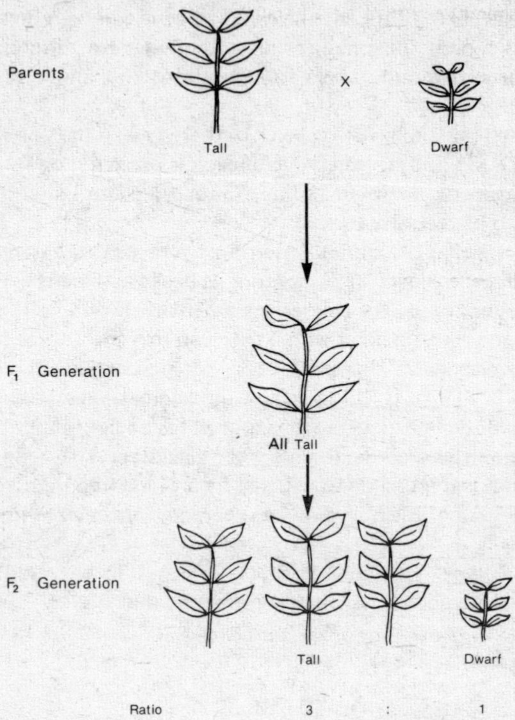

Fig. 13.3.

selfed or cross-pollinated the resulting F_2 generation included tall and dwarf plants in the ratio of 3 tall : 1 dwarf (see Fig. 13.3).

Inheritance of one factor (in this case height) is called **single factor** or **monohybrid inheritance**. If we represent these alleles with letters, we can see how they are inherited.

Let T = dominant allele for tallness.

Let t = recessive allele for dwarfness.

First consider the cross between a tall TT plant and a dwarf tt plant. Fig. 13.4 shows what gametes are formed and how they may combine.

You can see that all of the offspring (F_1) are tall heterozygotes.

When working these things out, the first step is to decide what alleles are carried by the gametes. Then sort out how they will combine. The method shown in Fig. 13.4 is one way. Two other methods are shown in Fig. 13.5.

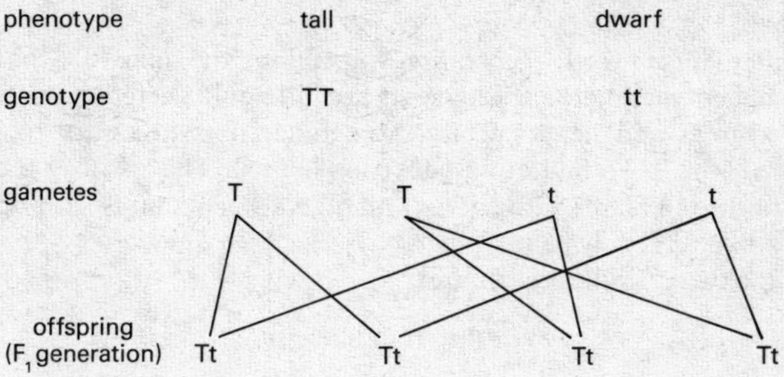

Fig. 13.4. *TT × tt*

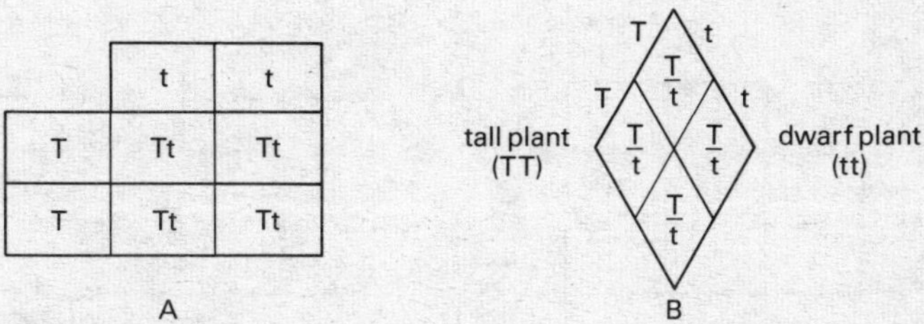

Fig. 13.5. *Different ways of working out crosses*

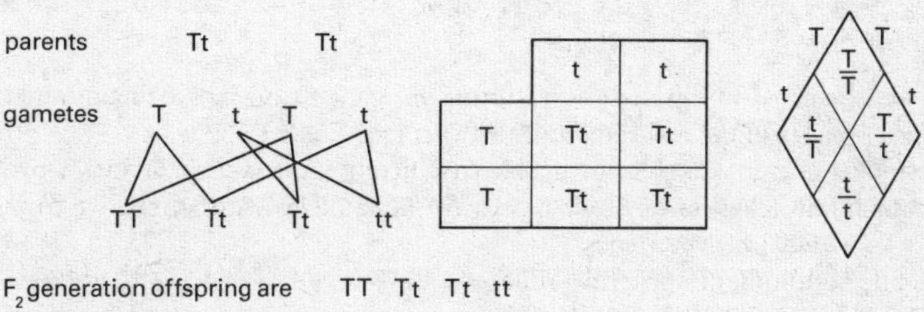

Fig. 13.6. *Tt × Tt*

The method shown in B is the one to be used by the Southern Examining Group.

You can experiment with these methods to see which you prefer.

Let's now see what happens when the F_1 are selfed. Fig. 13.6 shows what happens.

You can now see how the 3 : 1 ratio of tall to dwarf plants is produced.

The backcross

The tall plants in the F_2 generation will all look the same (they have the same phenotype) but their genotypes are different. There is a mixture of homozygotes and heterozygotes. Sometimes it is necessary to know which is which. To find out a **backcross** is used. This involves crossing each of the tall plants with a dwarf plant. A dwarf plant is used because its genotype can only be tt.

So the two possible crosses are:

TT × tt or Tt × tt

Each cross will give different ratios in the next generation (see Fig. 13.7).

parents TT x tt

	t	t
T	Tt	Tt
T	Tt	Tt

all tall

Tt x tt

	t	t
T	Tt	Tt
t	tt	tt

half tall, half dwarf

Fig. 13.7. *A backcross*

Now look at the example question below to see how to go about answering genetics questions.

Q. In the fruit fly Drosophila grey body colour is dominant to ebony body colour.

(a) What would be the result of crossing a homozygous grey-bodied fly with an ebony-coloured fly?

(b) What would be the result in the F_2 generation of crossing two of the F_1 flies?

(c) How would you determine the genotype of one of the F_2 grey-bodied flies?

Take it step by step

(a) First choose letters for your alleles. *Remember* to use the capital and small versions of the same letter.

Let G = grey body

Let g = ebony body

You are told the genotypes of the parent flies so the cross is GG × gg. The results of this cross are shown in Fig. 13.8.

parents GG x gg

	g	g
G	Gg	Gg
G	Gg	Gg

Fig. 13.8. *GG × gg*

All of the offspring are Gg, i.e. grey-bodied heterozygotes.

(b) You now know the genotypes of the F_1 flies from part (a) of the question.

Crossing two F_1 flies Gg × Gg (Fig. 13.9)

parents Gg x Gg

	G	g
G	GG	Gg
g	Gg	gg

Fig. 13.9. *Gg × Gg*

The F_2 generation contains a 3:1 ratio of grey- to ebony-bodied flies.

(c) To find the genotype of one of the F_2 grey-bodied flies you must do a backcross to an ebony fly. The two possible genotypes are GG and Gg. The two possible crosses are shown in Fig. 13.10.

parents GG x gg

	g	g
G	Gg	Gg
G	Gg	Gg

Gg x gg

	g	g
G	Gg	Gg
g	gg	gg

Fig. 13.10. *Using a backcross to determine the genotype of grey-bodied flies.*

If the grey fly was homozygous (GG) all of the offspring are grey-bodied. If the grey fly was heterozygous (Gg) half of the offspring are grey and half are ebony.

13.5. *Incomplete Dominance*

The examples looked at so far showed complete dominance. That is, if the dominant allele was present, that was the character that showed in the phenotype. The dominant completely masked the effects of the recessive allele.

In some cases, different alleles of a gene may both be expressed in the phenotype. An example is blood groups in humans.

The four blood groups A, B, AB and O are the phenotypes. They are shown below with their corresponding genotypes.

Phenotype	*Genotype*
A	AA or AO
B	BB or BO
AB	AB
O	OO

As you can see from the genotypes, O is recessive to both A and B but A and B are co-dominant. That is, if both are present the person will be AB. This phenomenon is known as **incomplete dominance**.

Definition: Incomplete dominance *is when more than one allele is expressed in the phenotype.*

13.6. *Variation*

Every species has a set of characteristics which all members possess in common. The characteristics distinguish one species from another. This type of variation between different groups of organisms is the basis on which the classification system works (see page 289).

Within a species, clearly, variation exists from one individual to another and from one generation to the next. Variation is produced by a combination of the individual's genotype and the influence of environmental factors on the expression of that genotype. For example, basic body shape and size can be modified by environmental factors such as quantity of food eaten, exercise, disease, etc. Full expression of the genotype can be prevented or allowed depending on environmental conditions.

Sources of genetic variation

1. **Mutation:** this is a sudden change in either the amount or structure of chromosomal material (DNA). Mutations are of two kinds. **Chromosome mutations** occur when the normal number of chromosomes is changed. For example, Down's syndrome (mongolism) in humans is a result of an individual having 47 chromosomes instead of 46. **Gene mutations** involve a change in the structure of DNA. Haemophilia (blood failing to clot) is caused by a gene mutation.

Mutations occur spontaneously at a very low rate, but this rate may be increased by certain mutagenic agents in the environment, e.g. certain chemicals and various types of radiation (atomic radiation, u.v. light, and X-rays).

Mutations result in the formation of new alleles (alternative forms of a gene). Most mutations are harmful and recessive. However, they do provide the main source of genetic variation within a population. This genetic variation is the raw material of evolution.

2. **Genetic reassortment:** this results from the rearrangement of genes through:

(a) crossing-over during meiosis (see page 251)

(b) the random assortment of chromosomes during meiosis (see page 251)

(c) the random fusion of gametes at fertilization.

Continuous and discontinuous variation

Genetically controlled characteristics generally fall into one of two categories: continuous or discontinuous variation.

1. **Continuous variation:** these are characteristics which show a continuous range of expression rather than clearly separated types. Height, weight and intelligence are examples of continuous variation. Such characteristics normally give a normal distribution curve. For example:

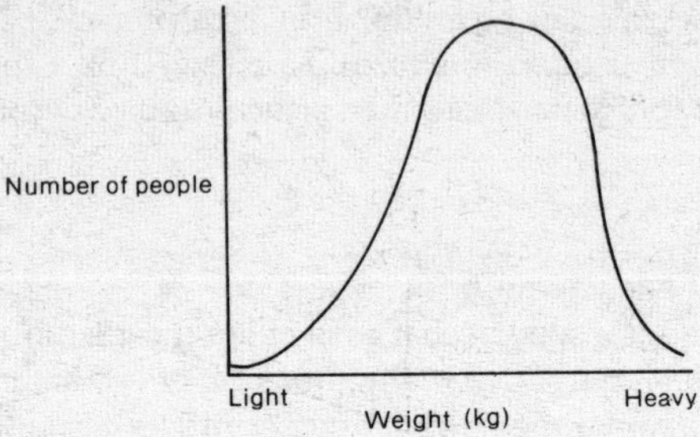

Characteristics which show a continuous range of variation:
(a) are usually controlled by several genes (multiple-factor inheritance).
(b) can be influenced by environmental factors.
(c) have a low heritability (the characteristics are not clearly passed on to offspring).

2. **Discontinuous variation:** these are characteristics which show a discontinuous range of expression, i.e. types are clearly separated without a range of intermediates. Eye colour in humans and tallness in sweet pea plants are examples. Characteristics which show a discontinuous range of variation:
(a) are usually controlled by one or two pairs of genes.
(b) cannot be changed by environmental factors.
(c) have a high heritability (the characteristics are clearly passed on to offspring).

13.7. *Sex Determination in Humans*

Of the 23 pairs of homologous chromosomes present in human cells, one pair determines the sex of the individual; these are the sex chromosomes. In the female, both chromosomes have the same appearance and are called X chromosomes. In the male, one chromosome is the X chromosome while the other is shorter and is called the Y chromosome. Gametes, being haploid, will only carry one sex chromosome.

It is the sperm which determines the sex of the offspring. The arrangement ensures that on average half the offspring are male and half female (see opposite). This method of sex determination is also found in other mammals.

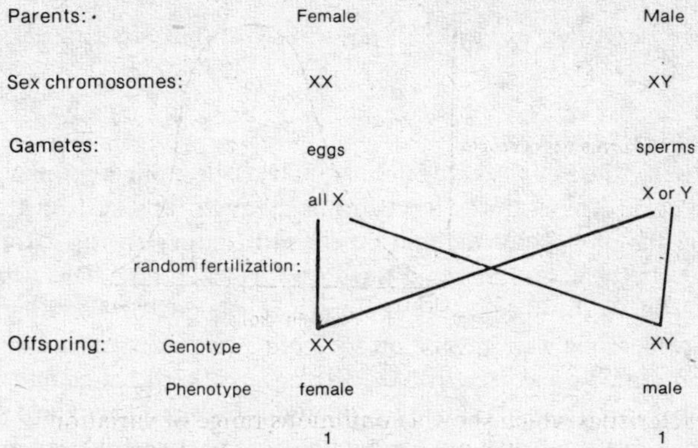

13.8. *Evolution and Natural Selection*

Until about the mid nineteenth century people believed that all species had been created and had remained unchanged. At about this time, however, scientists were beginning to think that there might be another explanation for the wide variety of living organisms. Their ideas became known as the theory of evolution.

The best known of these scientists is **Charles Darwin**. Darwin pointed out that all organisms need to be able to find food and shelter to survive and to produce young. Therefore, only organisms that were **adapted** to survive in their **environment** were able to produce offspring. Darwin had noticed that certain characteristics were passed from one generation to the next. He suggested that the characteristics that allowed an organism to survive were passed on to its offspring. He called this the **survival of the fittest**. Darwin's ideas are known as the **theory of natural selection**.

An example will help to show how natural selection works. The ancestors of today's giraffes did not have exceptionally long necks but some had longer necks than others. Those with the longer necks reached the higher leaves on trees and so had access to more food. When food was in short supply, those longer-necked animals with access to most food had a better chance of surviving and reproducing. If they reproduced, the genes for a slightly longer neck would be passed on to their offspring. These longer-necked offspring would be more likely to survive than their short-necked relatives, so the genes would be passed on to the next generation.

Gradually over many generations the necks of the population became longer, eventually resulting in today's familiar giraffe. It is important to remember that this happened only when there was an **advantage** to having a long neck.

There are other examples of natural selection favouring certain characteristics.

Industrial melanism

In Britain, the Peppered Moth has two varieties, one light in colour, the other dark. These moths feed during the night and rest on tree trunks during the day. Until the mid nineteenth century the light form of the moth was the commoner and the dark form was rare. This was because when resting during the day the moths depended on blending with the tree trunks they were resting on to avoid being seen and eaten by birds. In unpolluted areas tree trunks were covered with lichen and the light-coloured moths were difficult to see on this background. The dark forms, however, were conspicuous against the light background and easily spotted by birds.

As the amount of industry increased in certain areas, the lichen was killed by pollution and the tree trunks became blackened with soot. This meant that the darker moths were now less conspicuous while the lighter moths were easier to see. More of the lighter moths were eaten by birds, while greater numbers of the darker form survived to pass on their genes. By 1900 the darker form was common in the industrialized parts of Britain, while the light form became rare.

Since the introduction of the Clean Air Act in 1956 there is much less soot produced by factories and lichen is able to grow again in many areas. In such areas the light form of the moth now seems to be on the increase while the dark form is decreasing.

The dark colour of these moths is produced by a pigment called **melanin**, which is why this phenomenon is known as **melanism**.

Sickle cell anaemia

Sickle cell anaemia is a disease caused by a mutation in the gene that controls the structure of the haemoglobin molecule. The red blood cells of sufferers of this disease are liable to collapse into a sickle shape when exposed to low levels of oxygen, as found in the veins. We would expect this disease to be rare because people with it have reduced chances of survival. They would, therefore, have less likelihood of passing on this gene to the next generation.

In most parts of the world this is true but in some parts of Africa and Asia the gene is more common than would be expected. The reason for this is that people heterozygous for this gene have increased resistance to malaria. So, in areas where malaria is common and kills many people there is an advantage in carrying this recessive allele.

Resistance to antibiotics

Antibiotics, like penicillin, have been used to treat bacterial infections for many years. Every now and then a mutation occurs that gives a bacterium resistance to a particular antibiotic. This bacterium then has an advantage over other bacteria when they come into contact with that antibiotic. It is, therefore, able to survive and pass on this resistance to the next generation.

There are now many strains of bacteria which are resistant to a variety of antibiotics.

13.9. *Artificial Selection*

We continually seek to improve the crops we cultivate and the animals we domesticate. By careful breeding it is possible to change certain characters of animals and plants. For example, some plants can be bred to make them more tolerant to lack of water, or resistant to certain viral diseases. Cows may be bred to produce more lean meat and less fat, or to produce more milk. These types of artificial selection have been essential in keeping food production apace with population growth.

13.10. *Genetic Engineering*

Genetic engineering involves the transfer of a gene from one organism to another. For example, insulin-dependent diabetics need to receive regular treatment with the hormone insulin. Until recently the insulin was extracted from the pancreases of slaughtered animals. Genetic engineers have now been able to transfer the gene for insulin from a pancreas cell to a bacterium which is now able to make insulin. The bacteria are grown in large fermenters to provide a regular supply of the hormone. Other examples of the ways in which genetic engineering is being used are given below.

1. **Making hormones.** As well as insulin, human growth hormone is also being produced by genetically engineered bacteria. There is also the possibility of making a hormone that will increase milk production in dairy cattle.
2. **Making vaccines.** Some vaccines can now be made by genetic engineering, including a vaccine for leprosy.
3. **Interferon** is an anti-viral chemical that is produced in small quantities by our bodies. By means of genetic engineering, it is now being made in larger quantities for treatment of certain types of leukaemia.

4. **Making herbicides**. Many naturally occurring chemicals act as herbicides. By inserting the genes responsible for their production into bacteria these chemicals can be made in large quantities.
5. **Improved strains of crops.** It is also possible to transfer genes from one plant to another. In this way, plants can be given characters that will help them perform better under certain conditions.

Definition

Alleles

Key Words

Chromosomes
Genes
Diploid number
Haploid number
Homologous pairs
Mitosis
Meiosis
Replication
Crossing-over
Random assortment
Single-factor inheritance
F_1 generation
F_2 generation
Dominant
Recessive
Genotype
Phenotype
Homozygous
Heterozygous
Complete dominance
Incomplete dominance (co-dominance)
Chromosome mutations
Gene mutations
Continuous variation
Discontinuous variation
Selfing
Natural selection
Artificial selection
Melanism
Genetic engineering

Questions

1. (*a*) *What are the differences between haploid and diploid cells? Give* two *examples each of* (*i*) *haploid cells and* (*ii*) *diploid cells in mammals*

2. (*a*) *List the similarities and differences between meiosis and mitosis.*
(*b*) *Explain, using appropriate genetic symbols, the results from genetic crosses between:*
(*i*) *a homozygous tall plant and a homozygous dwarf plant*
(*ii*) *two heterozygous tall plants*
(*iii*) *a heterozygous tall plant and a homozygous dwarf plant. In all these crosses assume that tall is dominant to dwarf.*

3. Industrial melanism derives from the presence of a gene B *which leads to the development of a totally black colour in the adult of the Peppered Moth. This is dominant to the normal gene* b *which, in the homozygous state, results in a pale speckled form of the moth.*
A black moth was mated with a pale speckled one and approximately equal numbers of black and pale speckled offspring were obtained.
(a) Work out the genetics of this cross. What can you deduce about the genotypes of the parents?
(b) What would have been the result of mating two heterozygous moths together? Show all stages in your working.

4. In a genetic experiment a wild rabbit was mated with a tame white rabbit and it was found that all the offspring were brown, like the wild parent. These offspring were then mated with other white rabbits to produce a total of 216 offspring, 103 of which were white and 113 were brown. Explain fully the genetics of these crosses.

5. (a) Briefly distinguish between the terms
(i) discontinuous variation, and
(ii) continuous variation.
Give two *examples of each.*

14. Ecology and the Environment

Definition: Ecology *is the study of living organisms in relation to their surroundings.*

These surroundings are the **environment**, which is made up of living (**biotic**) and non-living (**abiotic**) components. The stable system formed by these components in a particular area is termed an **ecosystem**.

Definition: An ecosystem *is the living and non-living components of a region which interact to produce a stable system.*

Examples of ecosystems are a pond, a hedgerow and a beech wood. Ecosystems contain a number of **habitats** each with its own **community** of organisms.

Definition: A habitat *is a specific locality with a particular set of environmental conditions and its own* community *of organisms.* Put another way, a habitat is a place where an organism lives. In a pond ecosystem, the fringe (edge) of the pond is a habitat.

Definition: A community *is the collection of interacting species within a habitat.*

Within our fringe habitat, the common reed, *Phragmites*, the reedmace, *Typha*, the frog, *Rana*, and the mayfly, *Ephemera*, are members of the community.

14.1. *The Abiotic (Non-living) Environment*

The abiotic environment has both physical and chemical components. These conditions dictate which organisms may live in a given area and particularly affect the abundance of green plants, upon which animal populations depend. Some of the more important abiotic features are:

1. **Light:** solar radiation (sunlight) is the energy source for all life. It is required by green plants to make the organic substances on which all organisms depend (see page 48).

In flowering plants, day length coordinates events in the life cycle such as germination, flowering and fruit development. In mammals, day length triggers seasonal body changes like growth of winter coat, hibernation and reproductive behaviour.

2. **Temperature:** metabolic reactions, controlled as they are by enzymes, only operate efficiently within a narrow temperature range, generally 4–40 °C. Organisms, in general, choose habitats which will allow them to maintain an optimum temperature.

3. **Water** is the 'universal solvent' – the medium in which all life processes occur. It is also a transport medium and is a substrate for a number of important metabolic reactions, e.g. photosynthesis. Organisms living in dry environments must have a means of conserving water, e.g. a thick cuticle in plants, or efficient kidneys in mammals.

4. **Inorganic materials:** the availability of certain inorganic materials will affect plant and animal growth, either as nutrients, e.g. calcium, or poisons, e.g. heavy metals, or by producing unfavourable conditions like extreme pH (high acidity or alkalinity).

14.2. *Chemical Cycles*

Whereas energy flows through an ecosystem and is eventually lost, chemical resources are recycled. Three important chemical cycles are those of carbon, nitrogen and water.

Carbon cycle

Carbon is found in nature as atmospheric carbon dioxide (0.04 per cent of air), in other inorganic molecules, e.g. carbonate, within organic

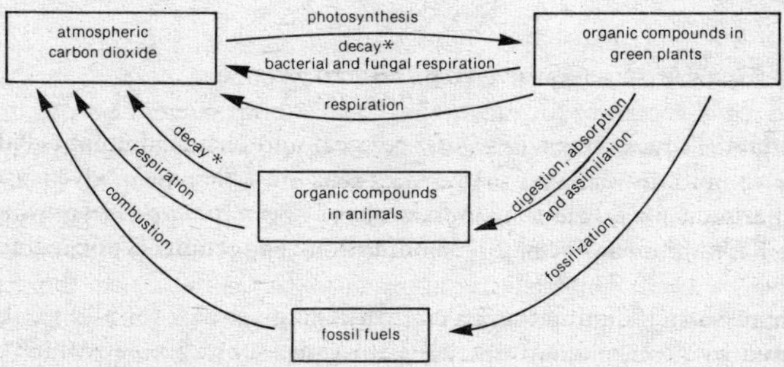

Fig. 14.1. *The carbon cycle*

* By bacteria and fungi.

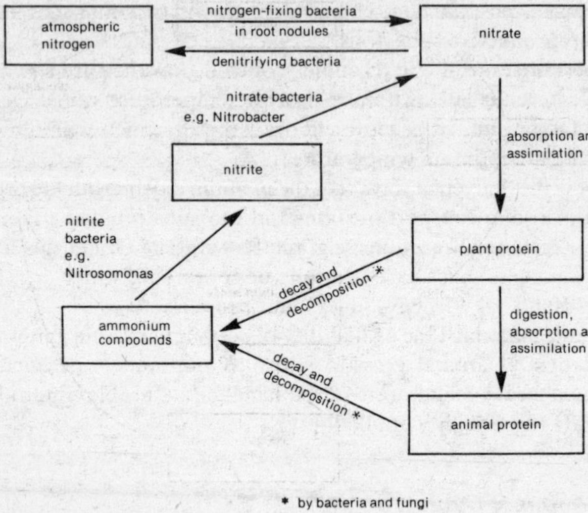

Fig. 14.2. *The nitrogen cycle*

molecules in living organisms, and within organic molecules outside living organisms, e.g. coal or man-made synthetic compounds.

Atmospheric carbon dioxide is the source of carbon for all green plants, and ultimately, for all animals. Photosynthesizing plants extract carbon and incorporate it into carbohydrates. Some of these carbohydrates are later converted into proteins and fats. Carbon is returned to the environment by respiration, decay or combustion (see Fig. 14.1).

Nitrogen cycle

Nitrogen is necessary for making proteins. About 80 per cent of atmospheric air is nitrogen. As a gas, it is inert (unreactive) and unavailable to most living organisms. However, nitrogen-fixing bacteria such as *Rhizobium* can use atmospheric nitrogen. They live symbiotically (see page 87) in the root swellings (nodules) of leguminous plants such as the pea or bean. The bacteria supply nitrogen (as nitrate) to the plant and receive sugar and protein in return. The plant uses the nitrogen for protein synthesis, animals obtaining their protein by eating plants. The nitrogen is recycled by decay of animal and plant material under the action of fungi and bacteria. **Nitrifying bacteria** convert ammonium compounds released by decay, first into nitrites and then into nitrates. De-

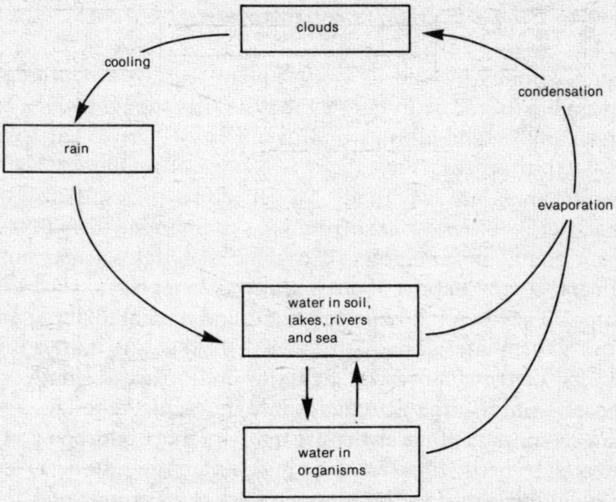

Fig. 14.3. *The water cycle*

nitrifying bacteria reduce the nitrate content of soil (and hence reduce the soil's nutrient value) by converting nitrates into nitrogen gas. The nitrogen cycle is summarized in Fig. 14.2.

Q. Soil bacteria play a vital role in the circulation of nitrogen in nature. Which one of the following bacterial processes is not *useful to flowering plants?*
A. the oxidation of ammonium compounds to nitrites
B. the oxidation of nitrites to nitrates
C. the fixation of nitrogen from the air
D. the conversion of nitrates to nitrogen gas
E. the conversion of animal and plant protein to ammonium compounds.

Water cycle

Water is circulated between the land, sea and atmospheric air (see Fig. 14.3). A small but significant portion passes through living organisms. Water is a raw material of photosynthesis (page 48) and is a source of hydrogen and oxygen for all living organisms. Plants and animals return water to the atmosphere in many ways, including excretion, transpiration (in plants), sweating (in mammals), and after death, by decay.

14.3. *The Biotic (Living) Environment*

Within a community and its environment there is a continual flow of energy and matter. Energy enters the system as sunlight which is trapped by green plants and during photosynthesis is converted to chemical energy within organic substances (see page 48). One way or another, these substances are eventually broken down in respiration (page 92) and the energy released is lost from the system in the form of heat.

Green plants are **producers** of organic material whereas animals are **consumers**. In any habitat there are also decomposers, such as bacteria and fungi. They break down dead plant and animal material and return it as inorganic material back to the soil. In this way matter is recycled: producers convert inorganic material into organic matter and decomposers convert organic remains into inorganic material.

In any community **food chains** are built up, the producers (green plants) being eaten by herbivores (plant-eating animals) and the herbivores being eaten by carnivores (meat-eating animals). An example of a food chain is shown in Fig. 14.4. Each stage of the chain is known as a **trophic level**.

Food chains in this simple form rarely exist. Each consumer usually has several food sources and in turn is preyed on by several organisms. This results in **food webs** developing in any community. Fig. 14.5 shows a food web which incorporates the food chain in Fig. 14.4.

Only 1 per cent of the light energy that reaches a plant is incorporated into plant tissues. As energy is passed along the food chain, about 90 per cent is lost between one level and the next (see Fig. 14.6). For example, of the energy required to maintain an individual rabbit, a large proportion is used in respiration to provide energy for movement, and to repair and maintain the body. This energy is eventually converted to heat. Only a fraction of the energy required to maintain a rabbit can be passed on

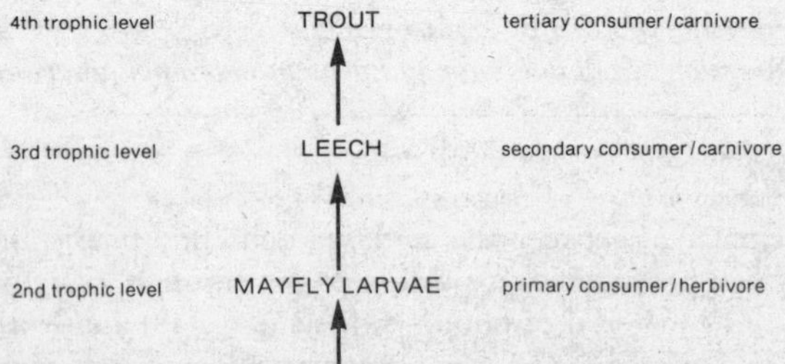

Fig. 14.4. *Food chain in a Welsh stream*

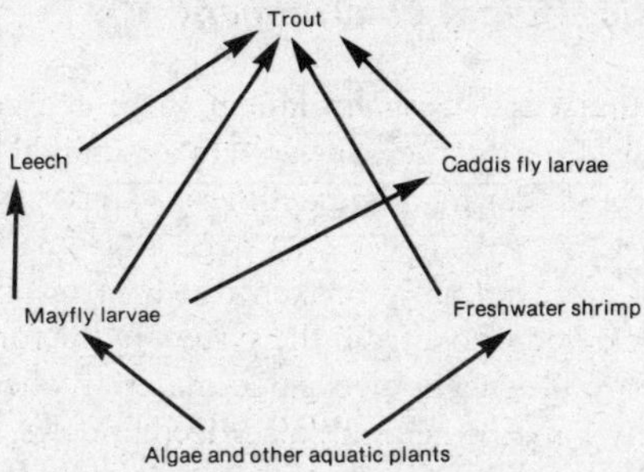

Fig. 14.5. *Part of a food web in a Welsh stream*

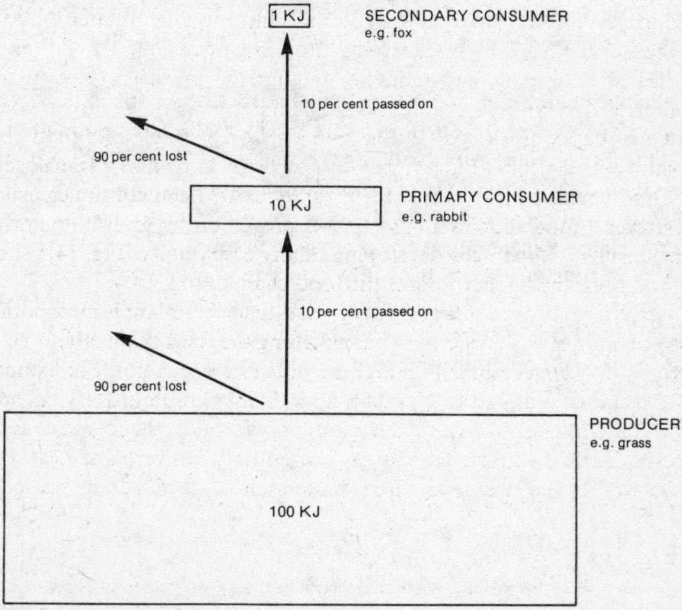

Fig. 14.6. *Energy flow through a food chain*
(values in boxes give energy content of tissue)

to a rabbit's predator, the fox. The fox can only obtain energy from digestible material in the rabbit.

This decrease in available energy in a food chain explains why higher trophic levels have a lower **biomass** (mass of living material); there is less energy available to support living material. It is this energy loss at each level which makes animal protein so expensive to produce. It explains

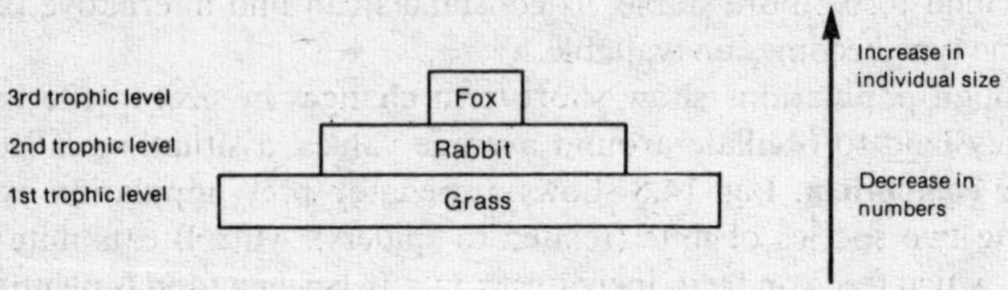

Fig. 14.7. *Pyramid of numbers*

why Third World countries tend to rely more on crops than livestock. It also explains why the number of links in a food chain rarely exceed five.

In general, the higher the trophic level in the food chain, the larger the size of the individual organism (after all, a predator must overpower its prey). A **pyramid of numbers** is produced (Fig. 14.7) as well as a pyramid of biomass.

Changed conditions in an ecosystem cause fluctuations in population sizes. For example, in our food web in Fig. 14.5, if the population of mayfly larvae were wiped out this would have serious repercussions throughout the community. The leech and caddis fly larvae would have to find an alternative source of food, or die. The freshwater shrimp might increase in numbers because there is more plant food available but, in turn, the trout will probably reduce their numbers by predation.

Simple ecosystems with few food chains, such as intensively farmed single-crop fields, tend to be seriously influenced by the removal of one species in the community. Complex ecosystems, such as tropical rain

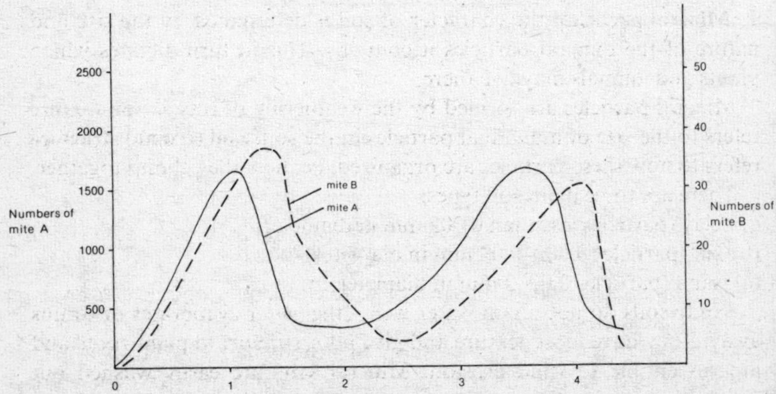

Fig. 14.8. *A predator–prey cycle for two species of mite*

forests, tend to be more stable, as consumers can find alternative prey when one type becomes unavailable.

Although populations show short-term changes in size, in the long term they tend to oscillate around average values, a situation called a **dynamic equilibrium**. Fig. 14.8 shows a predator–prey population cycle involving two species of mite (related to spiders). Mite B eats mite A. Mite A, which feeds on fruit, increases in numbers when food is plentiful. The population of mite B, which feeds on them, also increases. This results in a drop in the population of A, followed by a drop in the population of B, and the cycle then repeats itself. Notice, as you would expect, the numbers of mite A are higher than those of B.

14.4. *Soil*

Soil is a complex medium providing physical support and anchorage for plant roots and a supply of water and nutrients for plant growth.

Soil typically contains the following constituents:

1. mineral particles
2. water
3. air
4. dissolved mineral salts
5. humus
6. micro-organisms
7. soil animals.

1. **Mineral particles:** the character of soil is determined by the size and nature of the mineral particles it contains. This in turn dictates which plants and animals may live there.

Mineral particles are formed by the weathering of rocks. **Soil texture** refers to the size of individual particles in the soil. **Soil (crumb) structure** refers to how these particles are organized, i.e. how they clump together.

There are three main soil types:

(a) clay (particles less than 0.002 mm in diameter)
(b) silt (particles 0.002–0.02 mm in diameter)
(c) sand (particles 0.02–2 mm in diameter).

Sandy soils do not retain water well – the water evaporates or drains away. They have loose texture and give poor support to plant roots and are susceptible to wind erosion. Mineral salts are easily washed out (**leached**) from the soil.

Clay soils drain poorly, are heavy and are hard to work. The small particles are closely packed, hindering drainage and reducing aeration.

The soil (crumb) structure is governed by the soil texture (size of particles) and the proportion of dead organic matter (**humus**) present. The more organic matter present and the greater the mixture of different soil types, the more crumbly the soil.

A **loam** is a soil which contains a balanced mixture of sand and clay, and has a good humus content. It retains water but remains aerated and can be rich in mineral salts. Loams are usually excellent soils for growing plants.

2. **Soil water** exists as a thin film around soil particles and is held with an increasing force, the drier the soil becomes. The water content of a soil varies with soil type (see 1).

Available water is water that is available to be absorbed by plant roots. It includes capillary water (held to soils by capillary attraction) and drainage water.

Non-available water is bound more strongly to soil particles. Roots cannot exert a sufficient force to remove and absorb this water by osmosis (see page 137).

3. **Soil air:** air fills the spaces between soil particles except in a soil which is water-logged. Soils with large particles have more air spaces and drain rapidly. Soil air has more carbon dioxide and slightly less oxygen than atmospheric air because of aerobic respiration (page 92) by plant root tissues and other soil-dwelling organisms.

4. **Mineral salts** make up 0.02 per cent of the dry weight of soil. They originate mainly from the decomposition of organic matter (from plant and animal remains and excretory products). A range of these dissolved salts are essential for plant growth (see page 63). The balance of salts affects soil pH and therefore determines which plants can survive there.

5. **Humus** (dead organic matter) comprises the remains of dead organisms and their waste products. In soil used for growing crops humus rarely exceeds 6 per cent of the dry weight. Humus contributes to soil structure (see 1) and supplies nutrients and minerals as well as preventing their leaching out of the soil.

6. **Micro-organisms:** these include **saprophytic** bacteria and fungi (see pages 294 and 301). These break down organic compounds and humus to release soluble salts which are absorbed in solution by plant roots (page 136). **Nitrifying** bacteria convert ammonium compounds to nitrates (see page 270). **Nitrogen-fixing** bacteria (page 270) convert nitrogen gas to nitrate.

7. **Soil animals:** earthworms and beetles loosen the soil by burrowing, thereby increasing aeration and drainage. Earthworms contribute to soil fertility in a number of ways (see page 317).

NOTE. For certain examination boards you are required to know the experiments for determining soil composition, pH, and air, water or humus content. Check with your syllabus.

Q. (a) List the components of a fertile soil.
(b) Give four *ways in which earthworms improve soil fertility.*

Soil fertility

Agricultural practices aim to increase soil fertility.

1. **Manuring:** continued harvesting of crops removes soil nutrients which must be replaced. Natural (organic) fertilizers like manure (animal faeces) contain most, if not all, essential nutrients for plant growth, particularly nitrogen. They decay slowly and provide humus which maintains or improves soil structure.

Artificial (inorganic) fertilizers are quick-acting and can be selected to provide particular nutrients, e.g. nitrogen, phosphorus and potassium (NPK). They do not, however, improve soil structure and are easily leached into rivers and lakes, where they cause unwanted plant growth.

2. **Liming:** lime adds calcium to the soil, neutralizes acid and causes clay particles to cling together (**flocculate**), improving aeration and drainage.

3. **Ploughing** in spring breaks up soil clumps to produce a good seed bed. It also aerates the soil increasing microbial activity.

In autumn, ploughing opens up the soil to aid frost penetration. This breaks up the soil and helps to kill pests and weeds. Ploughing is also used to mix manure with the soil.

4. **Crop rotation**, such as the Norfolk Four Course (see Fig. 14.9).

Crop rotation achieves the following:

(a) It maintains the high nitrate content of soil. (NOTE. Clover is a leguminous plant, see page 270).

(b) It reduces pest number (most pests are specific for a particular crop and so do not survive when the crop is changed).

(c) It helps keep the soil weeded.

(d) It is economical on manure.

(e) Different crops draw mineral ions from different levels of the soil.

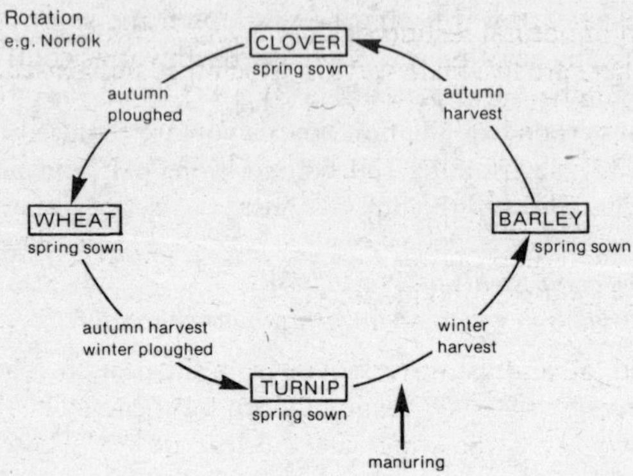

Fig. 14.9. *Cropping cycle*

Modern agriculture, through the use of artificial pesticides and fertilizers, need not rotate crops so rigorously, e.g. the intensive cereal cycle (see Fig. 14.10).

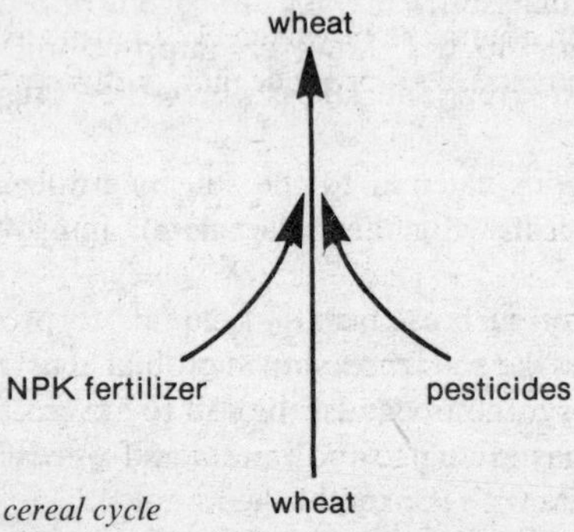

Fig. 14.10. *Intensive cereal cycle*

14.5. *Human Influences on the Environment*

Environment is a word used to describe our surroundings. The activities of humans have affected the environment in many ways. Over the last few hundred years the human population has increased dramatically (see page 247). With more people there are greater demands for food and housing. Meeting these demands has increased the effects we have had on our environment.

Exploitation of natural resources

Land use. There are five main ways in which the land is used:
Housing
Agriculture
Forestry
Mineral extraction – mining
Transport – roads and railways

Water is used in homes and industry. Fish are taken from rivers and seas for food.

Energy. Coal, oil and gas are removed from the ground. These are **finite** sources of energy; they will eventually run out. Energy is also obtained from sun, wind, moving water and nuclear power. These are **infinite** sources of energy.

Renewable resources are those which, if managed properly, can be maintained. Examples are food, vegetation and wildlife.

Non-renewable resources are those which will eventually run out. Examples are fossil fuels and minerals.

Pollution

A substance which enters the biosphere and harms living things is called a **pollutant**. Human activities produce many pollutants.

Air Pollution

1. **Burning fossil fuels.** Most air pollution is a result of burning coal, oil and gas, either in homes, factories or vehicles. Several pollutants are released when fossil fuels are burnt:

(a) **Soot** on leaves reduces the amount of light reaching the chlorophyll and limits photosynthesis. It also blocks the pores, or stomata, in the leaves, reducing gas exchange and water loss from the leaves.

The Clean Air Act of 1956 established smokeless zones in certain areas. This considerably reduced the amount of soot released into the atmosphere.

(b) **Sulphur dioxide** reacts with water vapour in the air to form sulphuric acid. This produces **acid rain** which damages plants, fish and buildings. Damage to the lungs from breathing sulphur dioxide increases the risk of lung disease in humans.

Most sulphur dioxide comes from power stations and could be reduced by the installation of filters in the chimneys of these buildings.

(c) **Carbon monoxide** from exhaust fumes combines with haemoglobin in the blood, reducing its capacity to carry oxygen.

It is possible to install filters in the exhaust pipes of cars which change the carbon monoxide to carbon dioxide.

(d) **Lead** from exhaust fumes accumulates in the body and can damage the brain, particularly in children.

Many countries are now introducing lead-free petrol and manufacturers are modifying car engines accordingly.

2. **Aerosols** release chemicals into the air which reduce the ozone layer and allow more ultra-violet light to reach the earth. It is feared that this could lead to an increase in skin cancer.

The use of pump-action dispensers instead of aerosols would help to alleviate these problems.

3. **Coal dust** from mines can cause a lung disease called **pneumoconiosis**. Coal miners wear masks to reduce the amount of coal dust that enters their lungs.

4. **Asbestos** causes a lung disease called **asbestosis**. Air leaving asbestos factories has to be filtered and workers have to wear protective clothing.

Steps are being taken to remove asbestos from buildings in Britain.

Water Pollution

1. **Sewage.** Untreated sewage entering water will provide food for bacteria. These bacteria will multiply, removing oxygen from the water and killing aquatic animals.

2. **Fertilizers** used on fields may drain into nearby rivers and streams. Nitrates in these fertilizers allow algae in the water to increase their numbers rapidly, causing **algal bloom**. This upsets the balance of nature in the water and damages the community of animals and plants.

Nitrates in drinking water can cause stomach cancer.

3. **Pesticides** used on fields may also drain into rivers and streams. Here they may enter food chains and accumulate in predators, particularly fish-eating birds. Fish are also very sensitive to certain pesticides.

4. **Heat.** Many factories take water from rivers and use it as a coolant. The heated water is then returned to the river. Hot water does not contain as much oxygen as cool water so there is less oxygen available to organisms in the river.

5. **Agricultural waste.** Factory farms produce large quantities of manure, usually more than can be used as fertilizer on the land. Excess manure is discharged into rivers, where it reduces the oxygen levels in the same way as untreated sewage.

6. **Industrial waste** may contain chemicals which are harmful in small quantities (e.g. lead, mercury, copper, cyanide). If these are released into rivers they may build up in fish and pass along the food chain.

7. **Oil.** When oil is discharged into the sea it floats on the water where it can damage birds and other organisms. Detergents used to break up oil slicks are also harmful to marine life.

Pollution from Pest Control

The effects of pesticides in water have already been mentioned. Pesticides can also cause damage on land. Some years ago crops were sprayed with DDT to kill insect pests. DDT did not break down very quickly so it entered food chains and accumulated in the bodies of certain animals, slowly poisoning them. Other pesticides have had the same effects.

DDT is now banned in Britain and most modern pesticides are broken down quickly in the soil.

Radiation

Low levels of radiation are natural and do not cause any harm. Very high levels from nuclear weapons or from major accidents at nuclear power stations can kill; lower levels from these sources can cause cancer and mutations in embryos.

Noise

This can be a form of pollution when there is too much of it. Persistent loud noise can cause deafness and depression. It has also been found to make people work less efficiently.

People working close to noisy machinery are required to wear ear protectors. Double glazing can help to reduce noise levels in houses near airports and motorways.

14.6. *Sewage Disposal*

Sewage contains a great many chemical compounds that provide food for bacteria. If sewage were poured straight into a river bacteria in the river would be able to multiply rapidly. Large numbers of bacteria would use up the oxygen in the water and fish would die. The level of organic materials in sewage must therefore be reduced before the sewage reaches the river. Sewage treatment also kills any harmful bacteria.

The first stage is **screening** to remove large objects. The sewage then goes to a **sedimentation tank** where large particles settle out to form a **sludge** at the bottom of the tank. The liquid left in the sedimentation tank is called **liquor** and may be treated in one of two ways.

1. Biological filtration

This is the older method of the two. The liquor is sprinkled over a filter bed of stones or coke. This filter bed contains protozoa and bacteria which feed on organic materials in the liquor as it trickles through the filter bed. At the end of filtration the liquid is emptied into a river.

2. Activated sludge process

The liquor from the sedimentation tank is mixed with a little sludge and pumped into long channels. Micro-organisms in the sludge digest organic materials in the liquor. To speed up the process air is bubbled through the liquid, mixing it and providing oxygen for the micro-organisms.

Treatment of sludge

Sludge from the sedimentation tank is transferred to a **digestion tank**. Here, large molecules are broken down into simpler substances. No air is allowed into the digestion tank so aerobic (oxygen-using) microbes are killed. Digestion is carried out by anaerobic (non oxygen-using) microbes. This creates less smell than aerobic digestion and produces methane which can be burnt as a fuel.

After digestion the sludge is dried on beds of sand and gravel. During this time the anaerobic bacteria in the sludge are killed by contact with oxygen.

Once dry, the sludge may be used as fertilizer. In some areas it is fermented with bacteria to produce more methane for use as fuel.

The process of sewage treatment is summarized in Fig. 14.11.

14.7. *Refuse Disposal*

Solid waste from homes and industry is a potential source of disease and great care must be taken in disposing of it.

Prior to collection refuse should be stored in containers with a lid, such as dustbins, or in plastic sacks which are sealed. These precautions prevent flies entering the rubbish and establishing breeding sites.

After collection the ways in which the refuse is dealt with varies from area to area.

Some local authorities put the rubbish on conveyor belts and remove any materials that can be **recycled**.

Inflammable refuse may then be **incinerated**. This is the safest way to

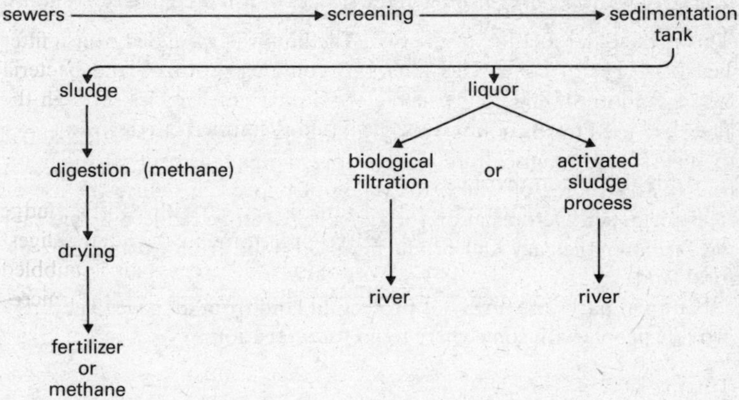

Fig. 14.11. *Sewage treatment*

deal with potentially infectious material such as used dressings from hospitals.

Remaining refuse is taken to waste ground where it is sprayed with disinfectant and covered with a layer of soil. Micro-organisms in the soil help to decompose the refuse. Each layer of refuse should not be more than 2 metres thick or decay will be too slow. These tips should be sprayed with insecticide to prevent flies breeding.

Sometimes burning or tipping is not possible. In these cases the refuse is packed tightly by heavy machinery and left to decompose slowly.

14.8. *Conservation of Natural Resources*

If we wish to continue using the natural resources that we exploit we must be able to conserve them.

Conservation of food resources

Farmers conserve much of our food resources by re-planting seeds after harvesting and also by maintaining breeding stocks of food animals.

Other sources of food should be similarly maintained, particularly by the fishing and whaling industries. This can be done by not killing animals faster than they can breed. International fishing agreements help to prevent the over-fishing of the species of fish used most commonly for food.

Conservation of trees

Trees are used for their wood and for making paper. Forests are cleared to make space for agriculture. Careful forestry management can maintain forests as a source of wood. Recycling of paper can reduce the use of trees for paper. Deforestation for agriculture can be reduced by increasing farming efficiency and producing more food from existing agricultural land.

National parks are a way of protecting land from damage. They also provide people with somewhere to go for recreation.

Conservation of wildlife

Many species have become extinct and many more are endangered because of the activities of humans. There are a number of ways to reduce this problem:

1. reduce pollution that damages natural habitats
2. avoid upsetting the balance of nature
3. avoid hunting animals unnecessarily, e.g. for sport, furs and ivory
4. introduce stricter laws against killing endangered species
5. Where possible, use substitutes for certain animal products, e.g. whale oil substitutes and synthetic fur and leather.

Conservation of minerals

We use enormous quantities of metals and the natural supplies of these will eventually run out. This process can be slowed down by recycling as much metal as possible, especially tin cans. The increasing variety of plastics has reduced the use of metal for some things.

Unfortunately, plastics are made from oil, which is also a non-renewable resource.

Conservation of fossil fuels

Oil, coal and gas are used for making energy. Coal and oil are also used to make fertilizers, dyes, medicines, plastics and many other products. Finding alternative methods of producing electricity will help to conserve fossil fuels.

There is a great deal of research into alternative forms of energy and the most promising types are listed below:

1. **nuclear power**
2. **solar power**
3. **wind power**
4. **water power**

Many countries already have nuclear power stations but there are considerable fears about the safety of these installations.

Recycling of plastics can help to reduce the demands on fossil fuels.

Definitions

Ecology
Ecosystem
Habitat
Community

Key Words

Carbon cycle	Pyramid of numbers	Acid rain
Nitrogen cycle	Dynamic equilibrium	Conservation
Nitrogen-fixing bacteria	Soil texture	Sulphur dioxide
Nitrifying bacteria	Soil (crumb) structure	Carbon monoxide
Denitrifying bacteria	Loam	Pesticide
Producers	Humus	Sedimentation
Consumers	Available water	Screening
Decomposers	Non-available water	Pollution
Food chain	Leaching	Chlorination
Trophic level	Saprophytic	Sewage
Food web	Manuring	Sludge
Biomass	Liming	Liquor
	Crop rotation	Filtration
	Environment	Coagulation

Questions

1. Questions (*a*) *to* (*e*) *refer to the numbered stages in the following diagram of part of the nitrogen cycle.*

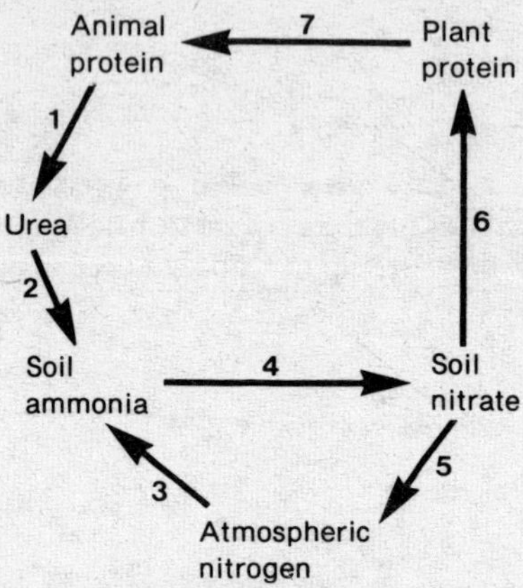

Which of the numbered stages represent the following processes?

(*a*) *Denitrification is*
A 2
B 3
C 4
D 5
E 7

(*b*) *Deamination is*
A 1
B 4
C 5
D 6
E 7

(*c*) *Nitrification is*
A 1
B 2
C 3
D 4
E 5

(*d*) *Nitrogen fixation is*
A 2
B 3
C 4
D 5
E 6

(*e*) *Decay is*
A 1
B 2
C 3
D 5
E 7

2. *The following diagram shows some of the feeding relationships of inhabitants of a British oak wood.*

(*a*) *Insert the following organisms or parts of an organism in the appropriate spaces in the food web.*

Oak tree leaves
Oak tree fruit (*acorns*)

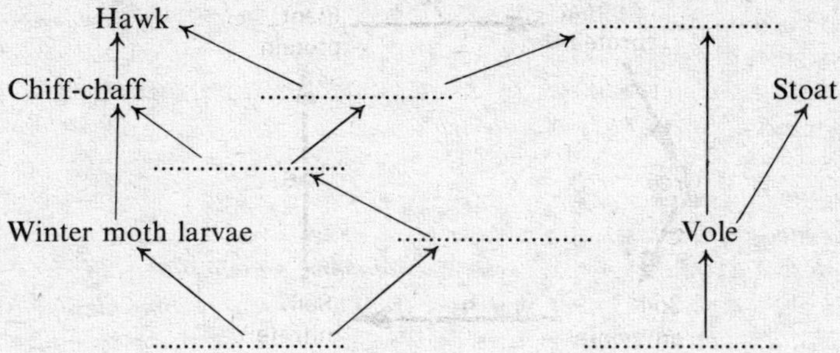

Blue-tit
Aphid
Ladybird
(*b*) (*i*) *When a stoat eats a vole energy is transferred to the stoat. Name* three *ways in which the stoat might use this energy.*
(*ii*) *Only a small proportion of the energy from the vole becomes incorporated into the stoat. What happens to the rest of the energy?*
(*c*) *Each autumn the oak trees shed their leaves. Explain how the nitrogen in the leaves is made available for the growth of trees in subsequent years.*

3. (*a*) *The diagram shows a simplified food web involving trout at different stages of its life history.*

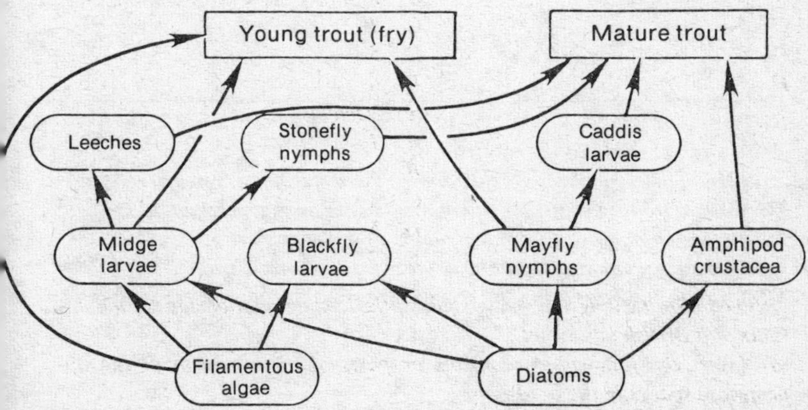

(*i*) *The ultimate source of energy for all the organisms in the food web is the sun. Describe* one *route by which the sun's energy may be made*

available for movement in trout. In your answer you should make reference to any processes important in the transfer of energy.
(ii) Using examples from the food web to illustrate your answer, explain what is meant by each of the following terms,
A producer
B primary consumer
C secondary consumer
(iii) Describe two *differences in the diets of young and mature trout.*
(iv) Suggest a possible effect on mature trout of a disease killing all mayfly nymphs. Explain your answer.

4. It has been suggested that lead from car exhausts that is washed into drains ends up in sewage sludge. What dangers might there be if this sludge were then spread on fields as fertilizer?

5. Table 14.1 shows the levels of sulphur dioxide in air and the incidence of bronchitis at different distances from a city centre.

Table 14.1

Miles from city centre	*Levels of sulphur dioxide in air* (μg/m³)	*% of population suffering from bronchitis*
0	85	16
2	62	13
4	53	9
6	25	5
8	18	3
10	4	1

(a) Draw a graph of the results.
(b) How does sulphur dioxide damage the respiratory system?
(c) What relationship do you notice between the percentage of people suffering from bronchitis and distance from the city centre?
(d) What explanation would you offer for this relationship?
(e) Name one other cause of bronchitis.

15. The Diversity of Life

In Chapters 5 to 12 we examined the seven characteristics of life – nutrition, respiration, excretion, sensitivity, growth, movement and reproduction – largely as applied to mammals and flowering plants. These are the most complex representatives of the animal and plant kingdoms. However, on all syllabuses you are required to have some knowledge of a range of organisms. A selection of these organisms is included in this chapter. Representative species from the main groups of organisms are briefly described. Only those details which are emphasized on syllabuses are included.

15.1. *The System of Classification*

The naming of living organisms follows a system of classification. This is really a way of organizing animals and plants into groups to help us describe and study them. There are three main advantages to the system of classification that is used:

1. It makes everyone talk the same language. Russian, Japanese and Australian biologists all understand that the scientific name *Canis familiaris* refers to the domestic dog.
2. It summarizes information. The name 'mammal' conjures up a picture of a hairy, highly developed animal, visible to the naked eye.
3. The system reflects evolutionary relationships between organisms. Several hundred years ago, organisms were grouped together according to seemingly related features. For example, earthworms, tapeworms and snakes were all classed as 'worms' because of their shape, and insects, birds and bats were placed in another group because they could fly. Nowadays, both animals and plants are grouped according to their evolutionary relationships. Snakes, for example, have a much more advanced body form than the tapeworm or earthworm. In terms of structure and physiology (function), snakes have much more in common with mammals, birds, fish and amphibians, and are included with these organisms in the group 'vertebrates' – animals with a vertebral column (backbone).

The groups that biologists use to classify organisms are known as **taxons**. The largest are **kingdoms**, of which there are two: the plant and animal kingdoms. These are broken down into progressively smaller taxons, the members of which have more and more features in common. The smallest groups are called **species**.

Definition: A species *is a group of organisms with a large number of structural and physiological (functional) features in common. The members of a species can interbreed to produce fertile offspring.*

Horses are an example of a species. Horses and asses are different, but closely related, species. They can interbreed, but the offspring so produced, called mules or hinnies, are infertile.

All species of organisms are allocated scientific names according to the **binomial** ('two name') system introduced by Linnaeus in the eighteenth century. All species are given two names in Latin:

1. The first name is the **genus** name and is written with a capital letter, e.g. *Homo* (man).
2. The second name is the **species** name and is written with a small letter, e.g. *sapiens* (wise).

The full name is either written in italics or is underlined, e.g. *Homo sapiens*.

Any organism can be precisely allocated within the classification system. Table 15.1 gives two examples. A mnemonic is provided to help you remember the list of taxons in the right order.

At this level of study, you need only be able to classify organisms to the level of phylum, or in some cases (arthropods, vertebrates and seed-bearing plants), to the level of class. However, check with your syllabus.

Table 15.1. *Use of the classification system*

Mnemonic	*Taxon*	*Examples*	
		Man	*Bulbous buttercup*
Keep	Kingdom	Animal	Plant
Plucking	Phylum (pl. phyla)	Chordata	Spermatophyta
Chickens	Class	Mammalia	Angiospermae
Or	Order	Primates	Ranales
Face	Family	Hominidae	Ranunculaceae
Getting	Genus (pl. genera)	*Homo*	*Ranunculus*
Sacked	Species	*sapiens*	*bulbosus*

15.2. *The Plant Kingdom*

The term division is sometimes used in place of phylum.

Phylum Bacteria: microscopic unicellular organisms with simple structure (see page 294).

Phylum Algae: simple aquatic green plants. Many are unicellular. Examples are *Chlamydomonas* (page 297) and *Spirogyra* (page 299).

Phylum Fungi: non-green plants (no chlorophyll). Either saprophytic or parasitic. Body consists of threads (hyphae). Examples are *Mucor* (page 304) and *Phytophthora* (page 302).

Phylum Bryophyta: mosses and liverworts. Small green land plants with simple leaves and root-like structures. Show alternation of generations. The gametophyte generation is dominant to the sporophyte generation.

Phylum Pteridophyta: ferns – green plants which have roots, stems and leaves but do not bear seeds. Show alternation of generations. The sporophyte generation is dominant. Example, *Dryopteris* (page 305).

Phylum Spermatophyta: green plants with roots, stems and leaves and which bear seeds. Have a well-disguised alternation of generations. The sporophyte generation is dominant (see page 213).

1. **Class Gymnospermae** (gymnosperms) – cone-bearing plants.
2. **Class Angiospermae** (angiosperms) – flowering plants. Examples, the buttercup, *Ranunculus* (page 313) and the chestnut tree, *Aesculus* (page 314).

15.3. *The Animal Kingdom*

It can be divided into those animals without a vertebral column, 'invertebrates', and those with a vertebral column, 'vertebrates'. All phyla from Protozoa to Echinodermata are invertebrates. Most members of the phylum Chordata are vertebrates.

Phylum Protozoa: single-celled animals. Example. *Amoeba* (page 315).

Phylum Coelenterata: aquatic. Body made up of two layers of cells surrounding a cavity – the enteron – whose only opening is the mouth.

Phylum Platyhelminthes (flatworms): small flattened worms which lack a body cavity. Many are parasitic.

Phylum Annelida (true worms or annelids): round worms with segmented bodies and through gut. Example, the earthworm, *Lumbricus* (page 317).

Phylum Arthropoda (arthropods): the largest animal phylum in terms

of species. Possess hard exoskeleton, jointed limbs and jointed appendages.
1. **Class Crustacea** (crustaceans).
2. **Class Arachnida** (arachnids).
3. **Class Myriapoda** (myriapods).
4. **Class Insecta** (insects).

Phylum Mollusca (molluscs): soft-bodied animals often with a shell. Examples, slugs, snails, oysters, octopus.

Phylum Echinodermata (echinoderms); marine 'spiny-skinned' animals based on a five-rayed plan. Examples, starfish, sea urchins.

Phylum Chordata (chordates): all those below are vertebrates.
1. **Class Pisces** (fish).
2. **Class Amphibia** (amphibians).
3. **Class Reptilla** (reptiles).
4. **Class Aves** (birds).
5. **Class Mammalia** (mammals).

15.4. *Viruses*

Viruses do not fit into either animal or plant kingdoms and are classified separately. Some biologists do not classify them as 'living' because they do not show *all* of the seven characteristics of life (see page 19). They can, however, reproduce.

Structure

They are **ultra-microscopic**, meaning that they cannot be seen using a light microscope, but only with the aid of an electron microscope. They are approximately 0.0001 mm in diameter which is about 10,000 times smaller than an *Amoeba*.

They come in a variety of geometric shapes, including spheres, rods and hexagons. A bacteriophage virus is shown in Fig. 15.1.

They lack a nucleus, cytoplasm and cell membrane and are therefore **non-cellular**. The body consists of a **protein coat** surrounding a strand of **nucleic acid** (DNA or RNA).

Mode of life

Viruses are **parasitic**. In order to reproduce they must invade a host cell and take over its metabolic machinery to make copies of themselves. In

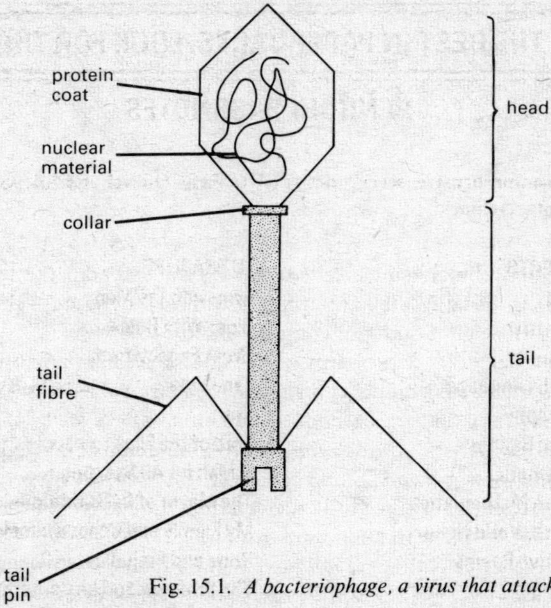

Fig. 15.1. *A bacteriophage, a virus that attacks bacteria*

so doing they burst and destroy the host cell. The damage viruses do to the host's tissues means they cause disease.

A particular virus normally has a specific host organism. Viruses are usually named after the disease they cause:

Name of virus	*Host*	*Mode of transmission*
Influenza (flu) virus	Man	Droplets (sneezing)
Poliomyelitis (polio) virus	Man	Water
Tobacco mosaic virus	Tobacco plant	Insects such as greenfly

Control

Outside the host cell, the virus is inert and some exist as crystals. They can be destroyed by heat, disinfectants or u.v. light. Inside the mammalian body, some viruses are attacked by antibodies in a similar way to bacteria (see page 131). In addition they are attacked by the chemical **interferon** (see page 132). Unlike bacteria, they are not affected by antibiotics.

Importance to humans

1. **Harmful:** they cause major diseases in man, cattle and food crops.
2. **Helpful:** some viruses can be used to control pest populations, e.g. myxomatosis virus to control rabbits.

15.5. *Bacteria*

They are unicellular (single-celled) but do not have all the features of normal cells. Usually they are grouped with plants because they have a cell wall (although it does not contain cellulose).

Habitat: bacteria are probably more widespread than any other group and live in most environments on earth, the reasons being:
1. they have a variety of methods of nutrition.
2. they can withstand a wide range of environmental conditions – different temperatures, pH, etc. Some bacteria form resistant spores.

Structure: bacteria have a diameter about 1/100 that of an *Amoeba*; they are just visible with a light microscope.

Unlike other cells they do *not* have membrane-bound organelles. For example, mitochondria are not present and the nuclear material floats 'free' in the cytoplasm.

Unlike plants they do *not* have:
1. a cell wall made of cellulose (instead it contains fat and protein)
2. a large vacuole.

Bacteria come in different shapes and sizes. Sometimes they group together in clumps or chains. They are classified according to shape:
1. spherical (coccus)
2. straight rod (bacillus)
3. curved rod (vibrio)
4. spiral (spirillum).

A generalized bacterium is shown in Fig. 15.2.

Nutrition: some bacteria are **autotrophic**, building up organic substances from simple inorganic substances, either using sunlight energy (photosynthesis) or the energy from the oxidation of inorganic substances, e.g. nitrite to nitrate (chemosynthesis). Photosynthetic bacteria do not contain chloroplasts; chlorophyll is scattered throughout the cytoplasm.

Most bacteria are **heterotrophic** and require 'ready-made' organic food. Of these, some are **saprophytic** and some are **parasitic**.

Saprophytic bacteria feed on dead organic material. Digestive enzymes

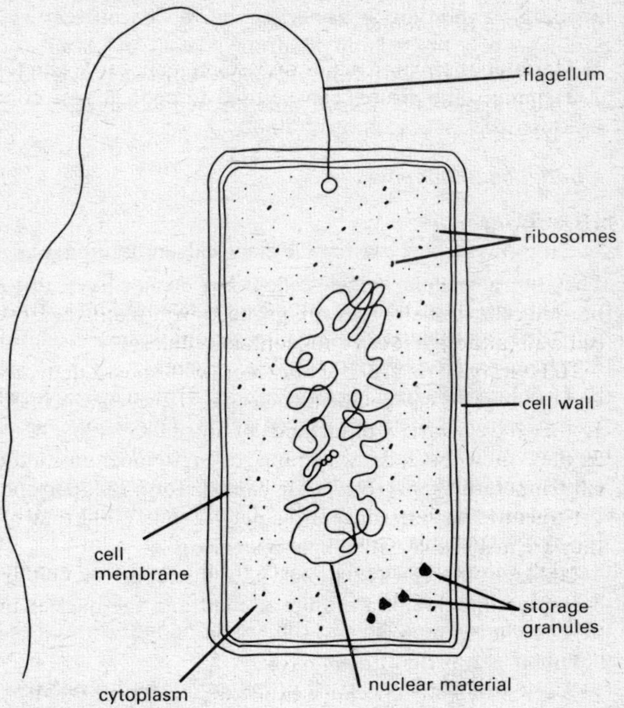

Fig. 15.2. *A generalized bacterium*

are secreted on to the substratum (surface on which the bacterium is found), food is digested **extracellularly** (*outside* the cell), and the products of digestion are absorbed across the cell wall.

Parasitic bacteria cause disease:

Name of disease	*Host*	*Mode of transmission*
Typhoid	Man	Eating or drinking contaminated food or water
Syphilis (venereal disease)	Man	Sexual contact
Tetanus	Man	Bacterial invasion of wounds
Pneumonia	Man	Water droplets in breath

Respiration: some aerobic, some anaerobic, some both.

Movement: some bacteria have flagella for locomotion (see Fig. 15.2).

Reproduction: asexual – by **binary fission** (dividing into two). Under favourable conditions a bacterium can divide once every 20 minutes, producing over one million offspring in less than 8 hours.

Some bacteria have a form of sexual reproduction called **conjugation**, whereby fragments of genetic material are exchanged.

Importance to humans

1. **Useful:**
(a) Saprophytic bacteria recycle chemicals in nature (see Figs. 14.1 and 14.2):
(i) **Nitrogen-fixers** convert atmospheric nitrogen into nitrates.
(ii) **Nitrifying bacteria** convert nitrogenous wastes into nitrates.
(iii) **Decay bacteria** break down dead organisms.
(b) **Cellulose-digesting** bacteria are used by herbivores which culture these bacteria in their gut (see page 87). This is an example of a symbiotic relationship.

NOTE. A **symbiosis** is an association between two organisms of different species where both partners derive benefit, frequently nutritional benefit. The rabbit benefits from the digestion of cellulose by the symbiotic bacteria; the same bacteria benefit from a continual supply of cellulose brought to them by the rabbit.

(c) **Vitamin-synthesizing** bacteria are found in the gut of man and other mammals.
(d) Certain bacteria are used in industrial food processes, e.g. making butter, cheese, yoghurt and vinegar.
(e) **Genetic engineering:** by genetic manipulation certain bacteria can be induced to make otherwise scarce and expensive chemicals, e.g. insulin.

2. **Harmful:**
(a) **Disease bacteria**, living in plants and animals, damage host tissue and excrete by-products which are harmful to the host.
(b) **Denitrifying bacteria** (see page 270) remove nitrogen from the soil, thus reducing its fertility.
(c) **Saprophytic bacteria** cause decay (putrefaction) of food.

Control

Bacteria require moisture, a suitable temperature and a food supply in order to survive. They are readily killed when exposed to u.v. light (found in sunlight).

Foods can be preserved by providing an environment unsuitable for bacteria:

1. **Drying** foods, e.g. peas, raisins, milk, meat.
2. **Salting** foods, e.g. fish, ham (high osmotic potential plasmolyses bacteria).
3. **Pickling** foods, e.g. onions (pH too acid for bacteria).
4. **Freezing** foods, e.g. meat (−20 °C suspends life).

Live bacteria can be prevented from being ingested by:

1. **Boiling** food – this kills most bacteria, but not bacterial spores.
2. **Pressure-cooking** ('sterilizing' or 'autoclaving') food – this kills all bacteria including spores.
3. **Pasteurizing** (heating and rapidly cooling) milk.
4. **Chlorinating** drinking water.
5. **General hygiene** and cleanliness about the house and use of **disinfectants** (chemicals which kill bacteria).

If bacteria invade the body and cause infection they can be counteracted by:

1. White blood cells which engulf bacteria (see page 132).
2. White blood cells which produce antibodies (see page 132).
3. Antibiotics (see page 132) administered to the sufferer, e.g. penicillin.
4. Injections of antibodies (see page 133) administered to the sufferer.
5. If the infection is on the body surface, e.g. sore throat, antiseptics (mild disinfectants) can be used.

15.6. *A Simple Plant:* Chlamydomonas

Phylum (Division): Algae.

Habitat: fresh water ponds and ditches. In large numbers it colours the water green.

Structure: about 0.02 mm long. It shows a mixture of plant and animal features (see Fig. 15.3).

Plant features:

1. a cellulose cell wall
2. chloroplast (for photosynthesis)
3. pyrenoid (region of starch formation).

Animal features:

1. flagella (for locomotion)
2. pigment spot (to detect light and move towards it – phototaxis)
3. contractile vacuole for osmoregulation.

Nutrition: autotrophic, by photosynthesis (see page 48).

Respiration: aerobic (see page 92).

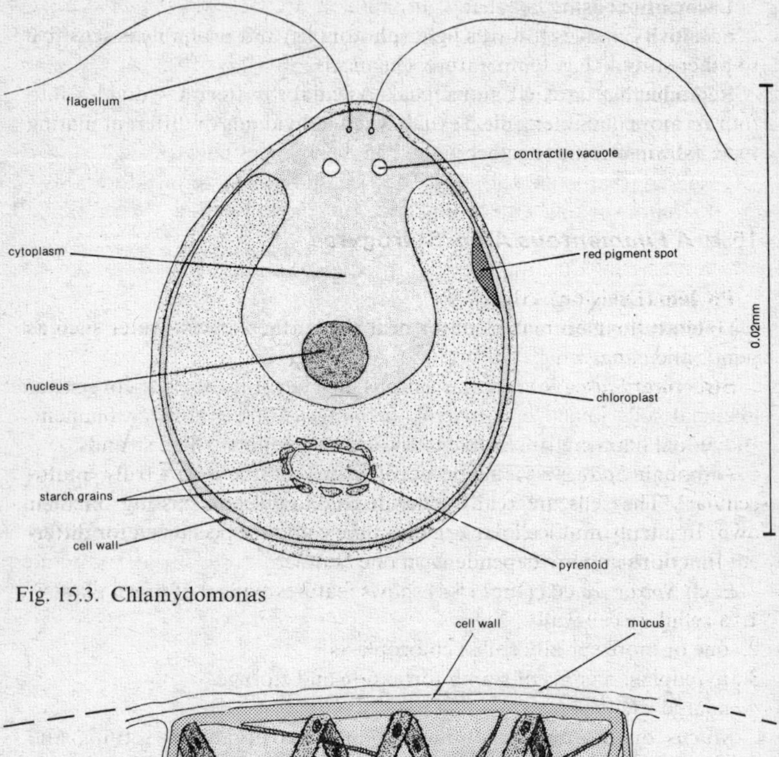

Fig. 15.3. Chlamydomonas

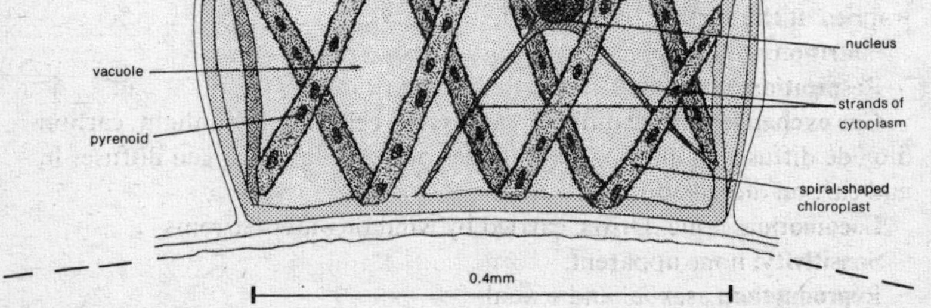

Fig. 15.4. *A* Spirogyra *cell*

Gas exchange: simple diffusion across cell wall. As with other green plants, in sunlight, carbon dioxide diffuses in and oxygen diffuses out. At night, oxygen diffuses in and carbon dioxide diffuses out (see page 62).

Locomotion: using flagella.

Sensitivity: moves towards light (phototaxis) and probably is sensitive to other stimuli, e.g. temperature, chemicals.

Reproduction: asexual and sexual. Asexual, by fission – divides into four or more daughter cells. Sexual, when individuals of different mating types (strains) come together.

15.7. *A Filamentous Alga:* Spirogyra

Phylum (Division): Algae.

Habitat: floats in masses on or near the surface of still water such as ponds and canals.

Structure: *Spirogyra* is a filamentous alga (see Fig. 15.4) It consists of identical cells joined end to end, forming a long strand or filament. Individual filaments are visible to the eye as green, hair-like strands.

Although *Spirogyra* is made up of many cells, it is not truly 'multicellular'. The cells are really individuals, capable of existing on their own. In a truly multicellular organism the cells are specialized for different functions and are dependent on one another.

Each *Spirogyra* cell (Fig. 15.4) shows features typical of green plants:

1. a cellulose cell wall
2. one or more spirally coiled chloroplasts
3. pyrenoids, regions of starch formation and storage
4. a large vacuole.

Mucus on the outside of the cell prevents organisms settling and growing on the plant. The mucus also prevents cells drying out when trapped at the surface of the pond.

Nutrition: autotrophic, by photosynthesis.

Respiration: aerobic.

Gas exchange: simple diffusion across the cell wall. In sunlight, carbon dioxide diffuses in and oxygen diffuses out. At night, oxygen diffuses in and carbon dioxide diffuses out.

Locomotion: none. Drifts, carried by wind or water currents.

Sensitivity: none apparent.

Reproduction: asexual and sexual.

Asexual by binary fission (splitting in two). This occurs in spring and early summer when conditions are favourable. A long strand (filament) of identical cells results. Fragmentation occurs if the strand breaks.

Sexual, by conjugation, between adjacent filaments (see Fig. 15.5). This occurs during periods of overcrowding and unfavourable conditions, such

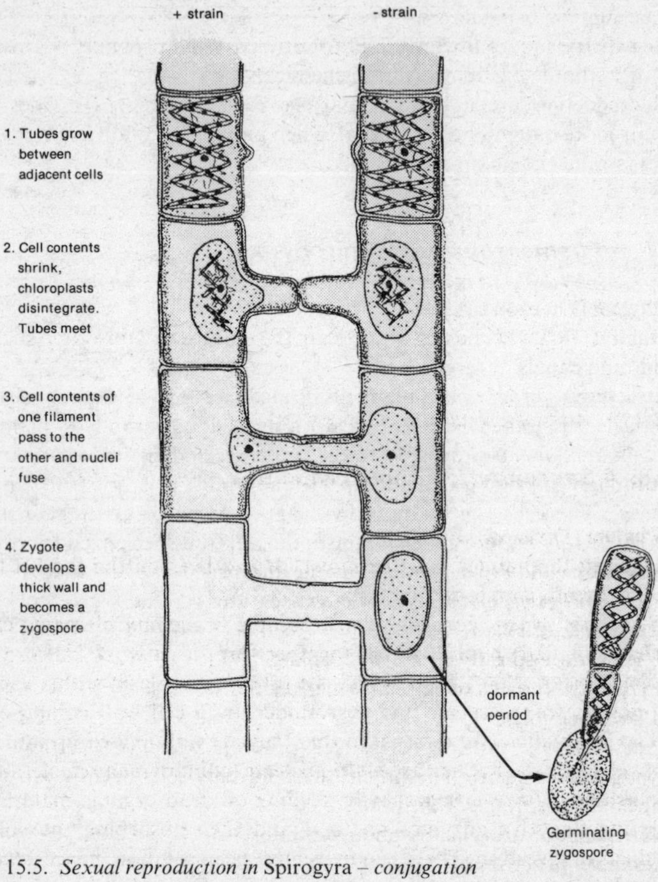

Fig. 15.5. *Sexual reproduction in* Spirogyra – *conjugation*

as occur in autumn. The end result is the production of a very resistant **zygospore** which is capable of overwintering. In conjugation, adjacent cells on two filaments form a conjugation tube between themselves. This acts as a bridge through which the contents of one cell enter the other. In effect, the contents of each cell becomes 'gametes' which, when they fuse, form a zygote. A hard capsule is then secreted, forming a zygospore. The zygospore falls to the bottom of the pond where it overwinters. The zygospore may be dispersed to other ponds in mud on the feet of animals. In spring, the zygospore splits, releasing a *Spirogyra* filament of one or two cells which by binary fission grows into a long filament.

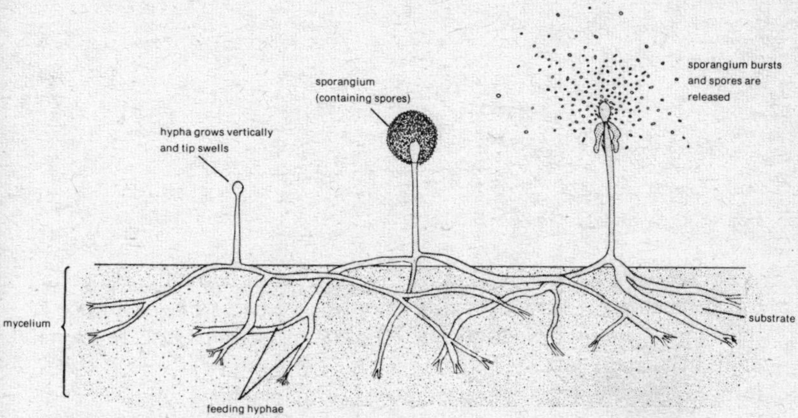

Fig. 15.6. *Asexual reproduction in* Mucor

15.8. *A Saprophytic Fungus:* Mucor

Phylum (Division): Fungi.

Habitat: the habitat *must* be moist. *Mucor* lives on the food it feeds on, e.g. bread, jam, football boots.

Structure: *Mucor* consists of microscopic branching filaments called **hyphae** (singular: hypha) which together form a network called a **mycelium** (see Fig. 15.6). Each hypha is a tube of cytoplasm with a vacuole running down the centre and surrounded by a cell wall composed of **chitin**. The hyphae are unusual in that they are not divided up into cells (non-septate); the hyphae are **multi-nucleate** (contain many nuclei).

Nutrition. *Mucor* is **saprophytic**, feeding on dead organic material by secreting digestive enzymes on to it and then absorbing the soluble products of digestion. These products must be in solution, hence *Mucor*'s dependence on a damp environment.

Respiration: aerobic.

Gas exchange: oxygen enters and carbon dioxide leaves by simple diffusion across the cell wall.

Reproduction: both asexual and sexual reproduction occur.

Asexual reproduction (see Fig. 15.6): vertical hyphae develop **sporangia** (fruiting bodies) which each produce hundreds of tiny spores. In certain species of *Mucor* these spores are very resistant to drying out (desiccation) and can survive adverse conditions. The spores are very light and can be carried long distances on the wind. They will germinate under suitable damp conditions.

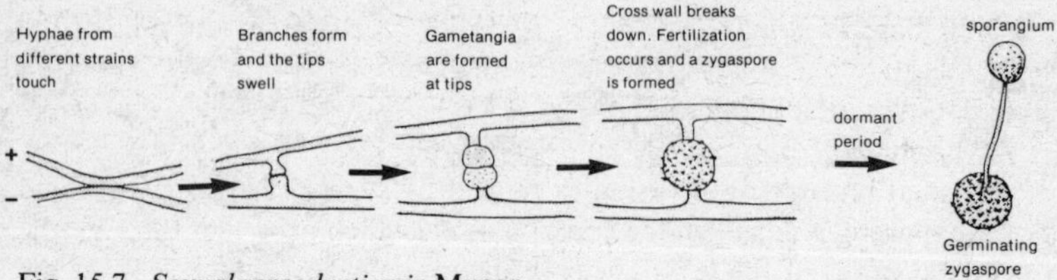

Fig. 15.7. *Sexual reproduction in* Mucor

Sexual reproduction (see Fig. 15.7): if hyphae of two different strains touch, **gametangia** (structures which produce gametes) may be formed at the point of contact. Nuclei (gametes) of the two strains fuse, to form a multi-nucleate zygote. As in *Spirogyra*, the zygote secretes a thick wall to form a **zygospore**. The zygospore remains dormant until favourable conditions exist, when it germinates to form a vertical hypha and sporangium which releases hundreds of spores.

15.9. *A Parasitic Fungus:* Phytophthora

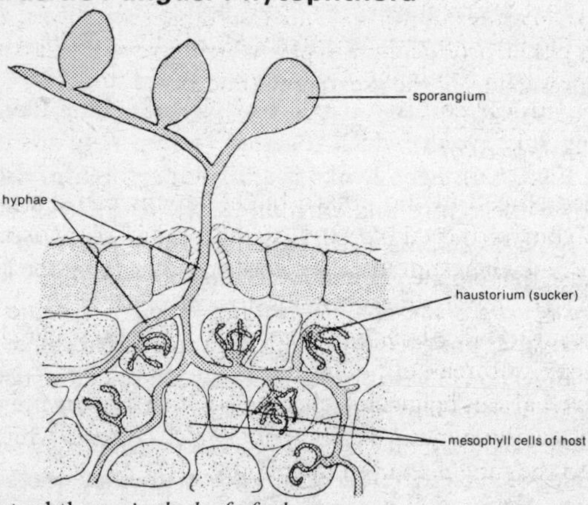

Fig. 15.8. Phytophthora, *in the leaf of a host*

Phylum (Division): Fungi.

Habitat: the fungus is a mould parasitic on the potato plant – it is known as the potato blight.

Structure: like *Mucor*, *Phytophthora* consists of non-septate multi-nucleate hyphae. In addition, it has specialized suckers (**haustoria**) which penetrate its host's cells and by which it absorbs food.

Nutrition: *Phytophthora* obtains its food directly from the cells of the host plant. Sucker-like haustoria secrete digestive enzymes and enter and break down mesophyll cells.

Reproduction: both asexual and sexual (rare).

Asexual reproduction is similar to that of *Mucor*. Spores are formed inside **sporangia** borne on the end of hyphae which grow out through the stomatal pores of the leaf. The sporangium is easily detached from the hypha and remains intact, being dispersed by air currents or washed away by rain. The spores are not as resistant as those of *Mucor* and must germinate within a few hours, on another plant, if they are to survive. A germinating hypha attaches to the surface of the host and develops a fine tube which bores through the cuticle into an epidermal cell. The tube swells and branching hyphae grow out between the host cells and produce feeding haustoria.

Sexual reproduction: two hyphae from different strains grow together, one forming distinct male (antheridium) and the other female (oogonium) gamete-producing structures. One male gamete will fertilize a single female gamete to produce a zygote which develops into an **oospore**, a resistant overwintering form. Under suitable conditions, the oospore germinates, either producing a hypha which grows into host tissues, or forming sporangia as in the asexual method.

Control of potato blight

Uncontrolled spread of the potato blight fungus can devastate potato crops. The tubers (the edible part of the potato plant, see page 207) become infected when sporangia are washed down from the leaves. After being harvested, these tubers rot in storage. The Irish famine in the nineteenth century, in which millions of people died, was mainly caused by a wide-scale outbreak of potato blight.

Nowadays, various **fungicides** (chemicals which destroy fungi) are available. These usually contain heavy metals, e.g. copper, which are more poisonous to the fungus than to the plant.

Importance of fungi

1. **Useful:**

(a) Saprophytic fungi play a major role in recycling dead organic matter (see Figs 14.1 and 14.2).

(b) Certain fungi produce antibiotics, e.g. *Penicillium* produces penicillin.

(c) Mushrooms and toadstools are fungi (albeit much larger than the moulds described here) and some are edible.

(d) Anaerobic respiration of yeasts is used to provide:
(i) alcohol, during fermentation to produce wine and other alcoholic beverages
(ii) carbon dioxide, to make the dough rise in bread.
2. **Harmful:**
(a) Decay fungi spoil food, e.g. *Penicillium* on jam, bread and cakes.
(b) Plant diseases such as potato blight are caused by parasitic fungi.
(c) Dry rot fungus destroys house timbers.

15.10. *Lichens*

Lichens are found growing on rocks and tree trunks. They are not a single organism but consist of an association between an alga and a fungus. There are many types of lichen, made up of different species of algae and fungi.

The fungus in these associations receives food which the alga makes by photosynthesis. What benefit the alga receives is not clear.

15.11. *A Moss:* Funaria

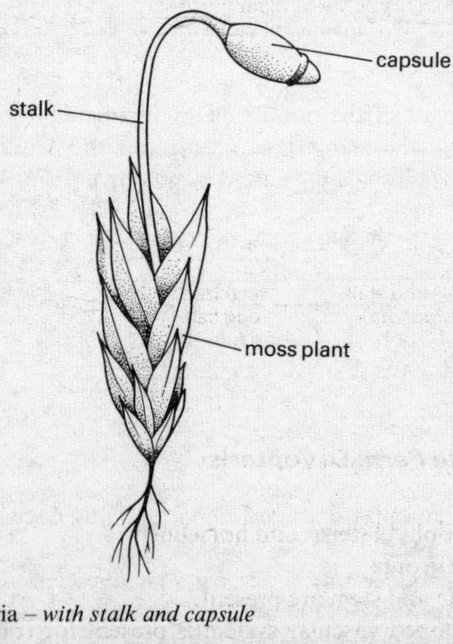

Fig. 15.9. Funaria – *with stalk and capsule*

Phylum: Bryophyta (mosses and liverworts)

Features of the group:

1. No distinct stem and leaves. Very simple root-like structures called **rhizoids**.
2. No specialized vascular system for uptake of water.
3. The life cycle shows an alternation of generations (see below).

Habitat: moist shady places. Common in woods both on the ground and on the trunks of trees.

Structure: see Fig. 15.9. Body is a flattened, branched structure called a **thallus**. Small root-like structures called **rhizoids** absorb water and mineral salts from the substrate. Following fertilization a capsule on a stalk develops from the moss plant (see Reproduction).

Nutrition: Mosses are **autotrophic**, feeding by photosynthesis.

Reproduction: the life cycle shows **alternation of generations**. Gametes produced by the moss plant (**gametophyte generation**) form a zygote which develops into the capsule and stalk shown in Fig. 15.9. This is the **sporophyte generation** and produces **spores**. Each spore gives rise to a branching filament from which buds grow. Each bud may grow into a new moss plant (see Fig. 15.10)

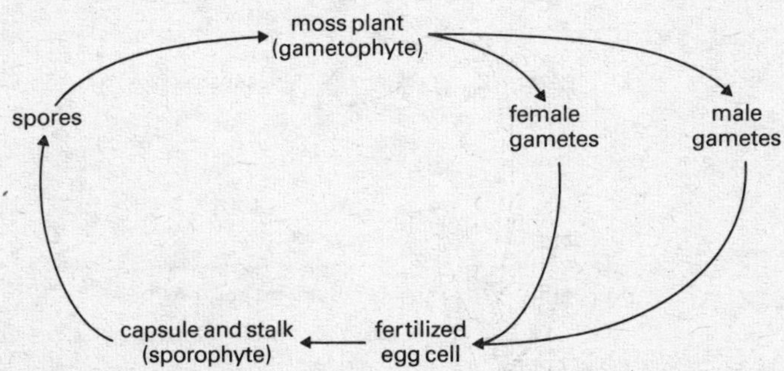

Fig. 15.10. *Moss life cycle*

15.12. *The Male Fern:* Dryopteris

Phylum: Pteridophyta (ferns and horsetails)

Features of the group:

1. Leaves, roots and stem are present.
2. A well-developed vascular system is present for transport of substances within the plant.

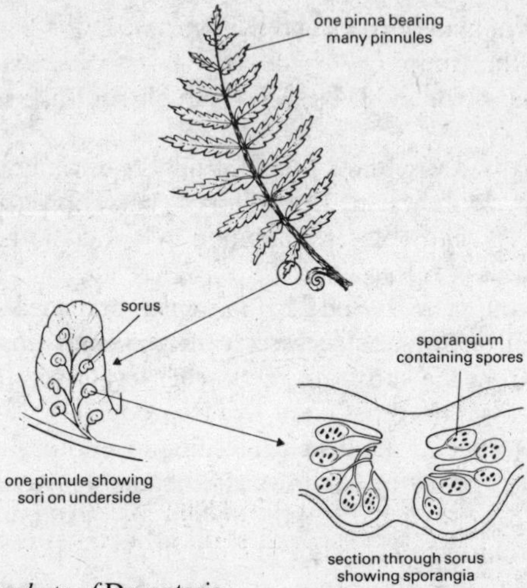

Fig. 15.11. *Sporophyte of* Dryopteris

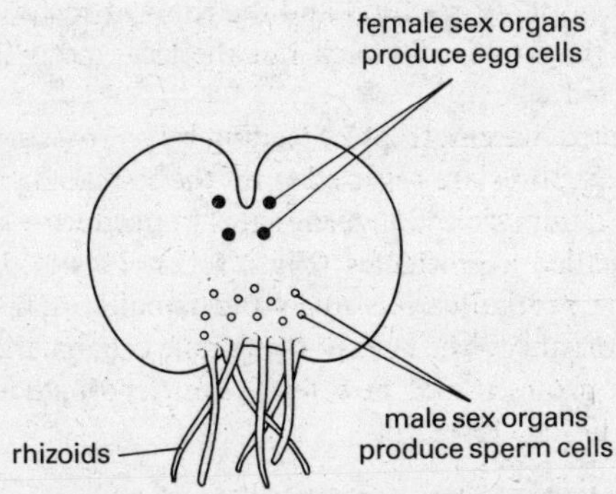

Fig. 15.12. *Underside of prothallus (gametophyte)*

3. Like mosses, ferns show an alternation of generations.
   Habitat: woods, hedgerows and open heathland.
   Structure: the sporophyte and gametophyte generations are separate. The sporophyte is the form by which most people recognize the fern (see Fig. 15.11).

The leaves (fronds) have a feather-like appearance and bear many

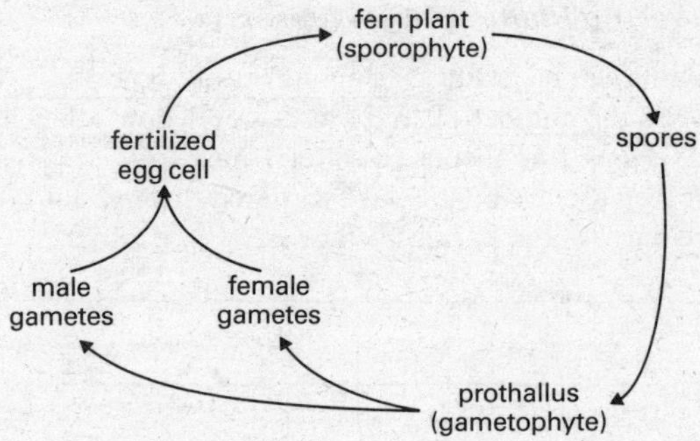

Fig. 15.13. *Life cycle of a fern*

pinnae, each of which is divided into **pinnules**. On the underside of the pinnules are **sori** (sing. **sorus**). Each sorus contains spore-forming **sporangia** (sing. **sporangium**).

The underground stem (**rhizome**) and roots anchor the fern to the ground. The rhizome stores food and the roots absorb water and minerals. In winter, the fronds die back but the rhizome with its food store remains protected.

Nutrition: Ferns are **autotrophic**, feeding by photosynthesis.

Reproduction: spores are released from the sporangia into the air. If a spore lands on damp ground it germinates to produce a small green leaf-like structure called a **prothallus** (Fig. 15.12). This is the gametophyte generation. The prothallus has male and female organs. Sperms swim from the male organs to the eggs in the female organs, the zygote formed by fertilization grows into a new fern plant (**sporophyte**). This process is summarized in Fig. 15.13.

Importance of alternation of generations

Terrestrial (land) plants such as mosses and ferns have developed this form of life cycle to overcome the problem of dispersion. As these plants do not move overcrowding would soon occur if all zygotes developed next to the parent plant. By incorporating a spore-producing phase in the life cycle, dispersion is ensured. Compare the dispersal mechanism in ferns with that in flowering plants.

15.13. *Flowering Plants (Angiosperms)*

Flowering plants can be grouped in a number of ways:

1. According to the number of seed leaves (cotyledons) they have:

(a) **Monocotyledons** (monocots for short) have one seed leaf. In the mature plant the vascular bundles are scattered throughout the stem and the leaves are narrow and parallel veined (see Fig. 15.14).

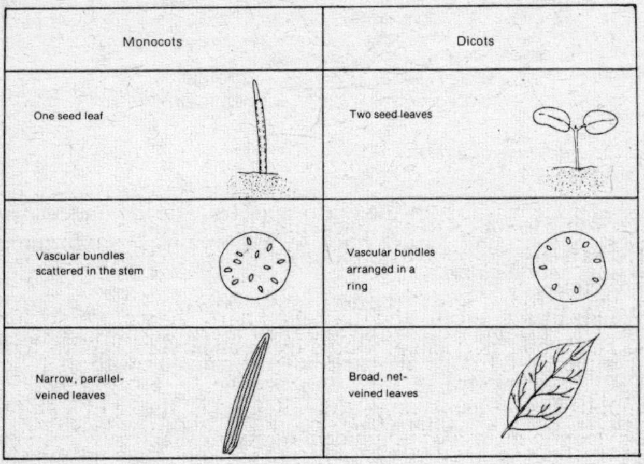

Fig. 15.14. *Monocotyledons and dicotyledons compared*

(b) **Dicotyledons** (dicots) have two seed leaves. This is the more common. In the mature plant the vascular bundles are arranged in a ring in the stem and the leaves are broad and net veined (see Fig. 15.14).

2. According to the duration of the life cycle:

(a) **Ephemerals** complete more than one life cycle in a year, overwintering as a seed. They are able to colonize new environments very rapidly. Many weeds are ephemerals, e.g. groundsel.

(b) **Annuals** complete one life cycle in a year, overwintering as a seed, e.g. broad bean.

(c) **Biennials** complete one life cycle every two years, overwintering as an underground food storage organ and then as a seed, e.g. carrot.

(d) **Perennials** complete one life cycle after several years. Woody perennials (see below) overwinter with trunk and branches remaining above ground. Herbaceous perennials (see below) overwinter as an underground food storage organ with (e.g. grass) or without (e.g. tulip) foliage above ground.

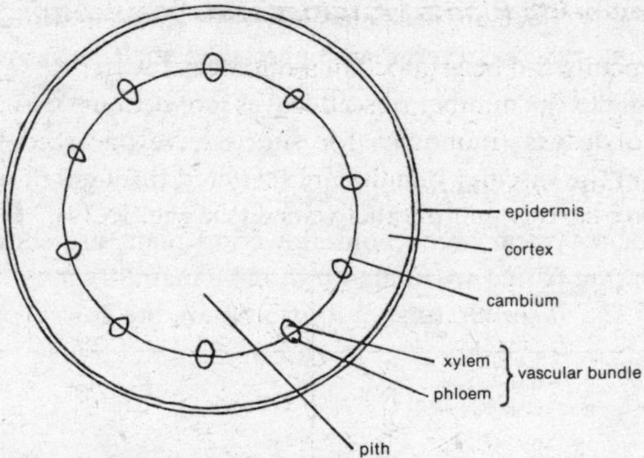

Fig. 15.15. *T.S. of a herbaceous dicot stem*

3. According to the degree of thickening in the stem:
(a) **Herbaceous** plants have unthickened soft green stems (see Fig. 15.15), e.g. buttercup, daisy.
(b) **Woody** plants, shrubs and trees, have hard, brown, woody stems and roots. The 'wood' is xylem produced by **secondary** thickening; it provides the stem with support to bear the weight of branches and leaves. A continuous ring of meristematic (dividing) cells, the cambium, produces extra xylem on its inside and phloem on its outside (see Fig. 15.16). You will notice the xylem and phloem are not restricted to vascular bundles but are in a continuous ring or rings. Xylem forms growth rings, each one corresponding to a year's growth. A cork cambium below the epidermis gives rise to a layer of cork which is impermeable to water.

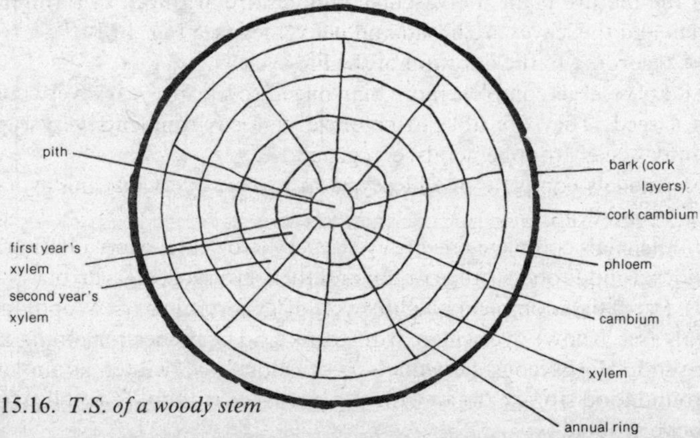

Fig. 15.16. *T.S. of a woody stem*

Woody plants which are **deciduous** lose their leaves in winter, e.g. oak, horse chestnut, whereas **evergreen** plants shed their leaves at irregular intervals, e.g. holly.

Structure of flowering plants

The parts of a typical herbaceous flowering plant are shown in Fig. 15.17. Flowering plants are made up of *three* main organs – stem, roots and leaves – and at certain times a fourth organ, the flower (page 213).

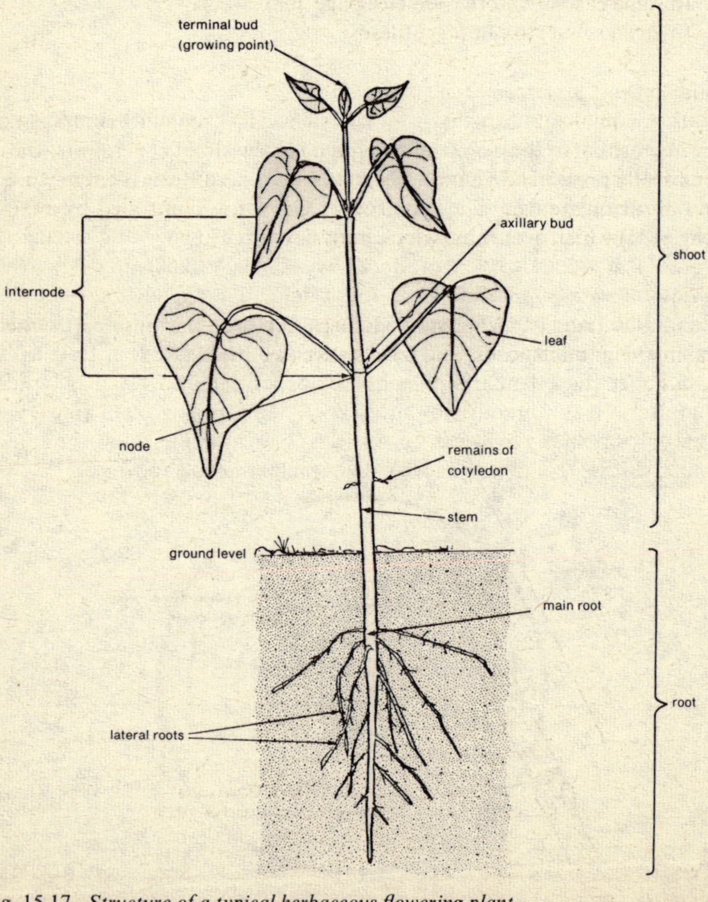

Fig. 15.17. *Structure of a typical herbaceous flowering plant*

The stem

The above-ground part of the plant is the shoot. The stem is the part of the shoot which bears the leaves and buds; it develops from the seed plumule (page 218).

The main functions of the stem are:

1. To hold the leaves aloft for photosynthesis (see page 48).
2. To hold the flowers aloft for pollination and seed dispersal (see pages 215 and 219).
3. To transport substances between the roots and leaves (see page 143).
4. In some cases, to store food (see page 208).
5. In green stems, to photosynthesize.

Buds

Buds are small outgrowths from the stem which contain compressed, dormant shoots capable of developing into new shoots or flowers; these are usually protected by bud scales. Buds are generally dormant in winter and in spring are stimulated to growth. Growth is controlled by auxins (page 185) which determine which buds develop when.

Twigs

Deciduous trees and shrubs shed their leaves at the onset of winter, leaving twigs of characteristic appearance (see Fig. 15.18).

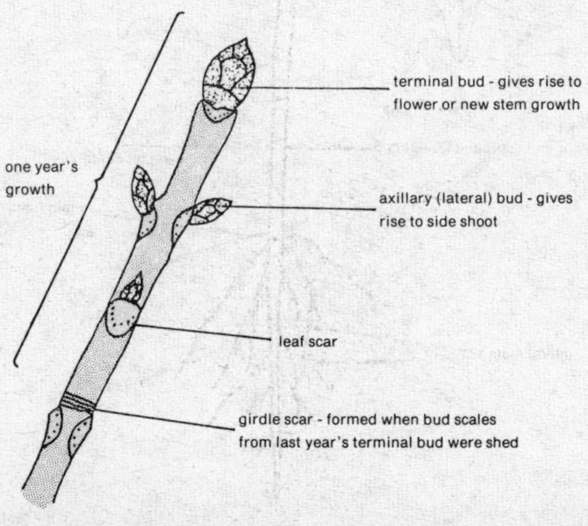

Fig. 15.18. *Horse chestnut twig in winter*

Roots

The main functions of roots are:

1. To anchor the plant in the ground.
2. To absorb water and dissolved mineral salts from the soil (see page 136).
3. To transport water and mineral salts to the stem (see page 136).
4. In some cases, to store food, e.g. the root tubers of the dahlia.

The structure of a young root tip is shown in Fig. 15.19.

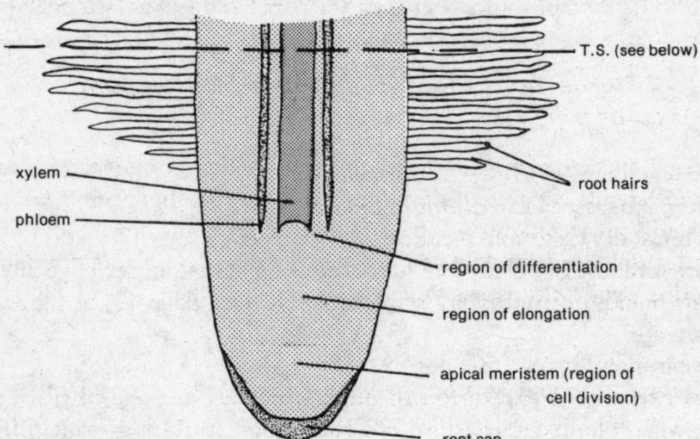

L.S. of root tip

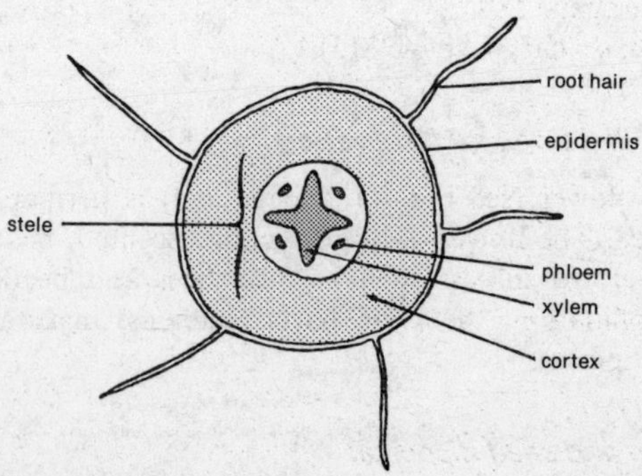

T.S. of root hair region

Fig. 15.19.

Leaves

These are attached to the stem at **nodes**.

The main functions of leaves are:

1. To carry out photosynthesis (see page 48).
2. To carry out gas exchange for respiration and photosynthesis (see page 113).
3. Transpiration (see page 138).
4. Amino acid synthesis (see page 52).
5. In some cases, to store food, e.g. the bulb of an onion (page 208).

15.14. *A Herbaceous Plant: the Bulbous Buttercup,* Ranunculus bulbosus

Phylum (division): Spermatophyta.

Class: Angiospermae (flowering plants).

Habitat: dry fields and meadows.

Structure: the buttercup is a herbaceous dicotyledon (see page 309).

Nutrition: autotrophic, by photosynthesis (see page 48) in the leaves and stem.

Respiration: aerobic (see page 92).

Gas exchange: by diffusion through stomata (see page 113). In sunlight, when photosynthesis exceeds respiration, carbon dioxide diffuses in and oxygen diffuses out. At night, respiration alone occurs, and oxygen diffuses in and carbon dioxide diffuses out.

Transport: see page 134.

Excretion: see page 148.

Sensitivity: see page 180.

Reproduction: sexual (see page 213).

The flower, pollination and fertilization

The buttercup flower (see Fig. 11.2, page 214) is hermaphrodite and insect-pollinated. The flower is bright yellow, scented, has nectar, and attracts a variety of insects such as flies, bees and beetles. To help prevent self-pollination, the male parts (stamens) mature before the female parts (stigmas).

Fruit formation and seed dispersal

After pollination and fertilization, each fertilized ovule forms a seed surrounded by a leathery fruit case. The seeds stay attached to the receptacle which breaks off to form a light bunch of fruits which is dispersed by the wind.

Germination

Germination is **hypogeal**, the cotyledons remaining below ground (see page 223).

This particular species of buttercup is **perennial** (see page 309). It overwinters as an underground food storage organ.

15.15. *A Deciduous Tree: the Horse Chestnut,* Aesculus hippocastanum

Phylum (division): Spermatophyta.

Class: Angiospermae (flowering plants).

Habitat: municipal parks and edge of fields.

Structure: see Fig. 15.20. The horse chestnut is a woody dicotyledon (see page 309).

Nutrition: autotrophic, by photosynthesis (see page 48) in the leaves. The stem is covered in bark and does not photosynthesize.

Respiration: aerobic (see page 92).

Gas exchange: by diffusion through stomata and lenticels (page 113).

Reproduction: sexual (see page 213).

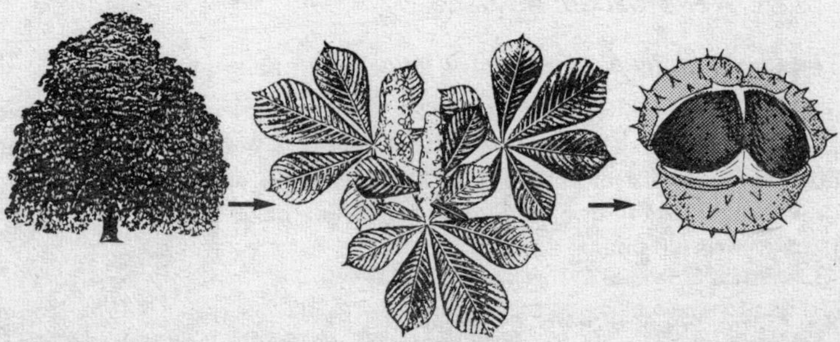

Fig. 15.20. *The horse chestnut*

The flower, pollination and fertilization

The flowers open in May. The 'candle-like' inflorescence (flower arrangement) consists of many single flowers attached to an upright shoot (see Fig. 15.20). The flowers are insect-pollinated. Bees are attracted by the white petals with orange bases and nectaries. Both male only and hermaphrodite flowers may be found in a single inflorescence. Within a hermaphrodite flower, the female parts (stigmas) mature before the male parts (stamens) to minimize self-pollination.

Fruit formation and seed dispersal

After pollination and fertilization, each fertilized ovule forms a very large seed (a conker) surrounded by a spiny, leathery fruit case (see Fig. 15.20).

The seed may germinate where it falls or may be dispersed by animals either carrying or eating the seed. Large mammals such as deer and cattle may carry fruit cases trapped in their fur or may eat the seed which occasionally passes through the digestive system unharmed. Squirrels burying conkers is a common way in which seeding takes place.

Germination

Germination is **hypogeal**, the cotyledons remaining below ground (see page 223).

15.16. *A Protozoan:* Amoeba

Phylum: Protozoa.

Habitat: still or slow-moving fresh water.

Structure: see Fig. 15.21. Up to 1 mm in diameter. Irregular shape, constantly changing. A single-celled (unicellular) animal.

Animal features:

1. no cell wall.
2. contractile vacuole (for osmoregulation).
3. food vacuoles (see nutrition below).

Nutrition: holozoic. Feeds on bacteria (page 294), algae such as *Chlamydomonas* (page 297) and other protozoa. Ingests its prey by **phagocytosis** (Fig. 15.22) using pseudopodia. Digestive enzymes are secreted into the

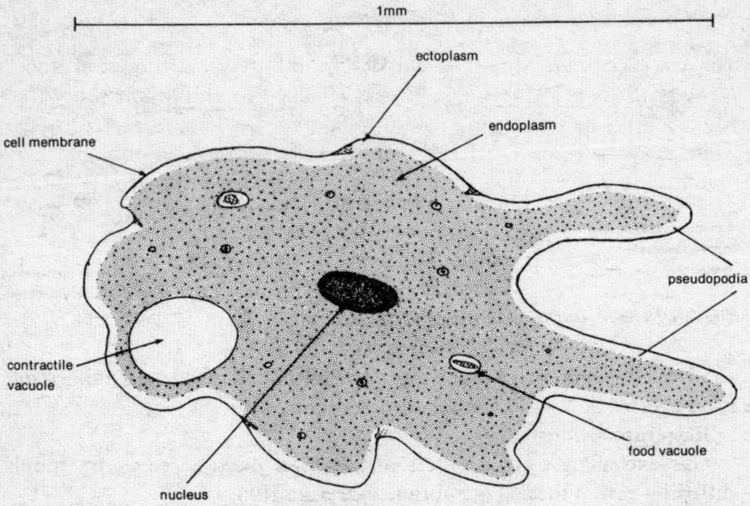

Fig. 15.21. Amoeba proteus

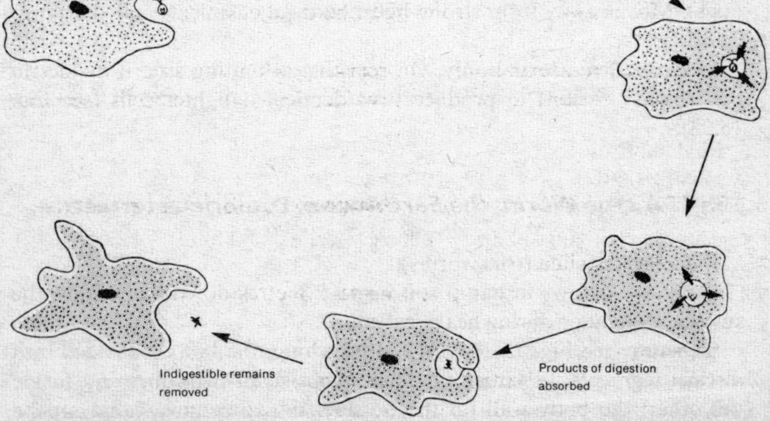

Fig. 15.22. *Feeding in* Amoeba

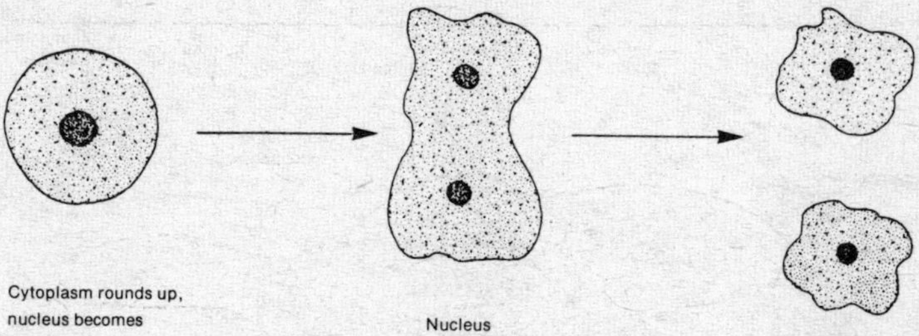

Fig. 15.23. *Binary fission in* Amoeba proteus

food vacuole, the soluble products of digestion absorbed, and indigestible food egested.

Respiration: aerobic.

Gas exchange: oxygen enters and carbon dioxide leaves by simple diffusion across the cell membrane (see page 101).

Excretion and osmoregulation: soluble nitrogenous wastes diffuse out across the cell membrane. Excess water which enters by osmosis (page 40) is pumped out using the contractile vacuole (see page 150).

Movement: *Amoeba* moves by pushing out pseudopodia ('false feet') into which it flows. Liquid cytoplasm (plasmasol) is thought to flow through a funnel of jelly-like cytoplasm (plasmagel) to form and extend the pseudopodium.

Sensitivity: follows the trail of chemicals released by its prey (chemotaxis). Moves away from strong light, harmful chemicals and sharp objects.

Reproduction: asexual only. On reaching a limiting size, it divides in two (binary fission) to produce two identical daughter cells (see Fig. 15.23).

15.17. *A True Worm: the Earthworm,* Lumbricus terrestris

Phylum: Annelida (true worms).

Habitat: burrows in damp soil up to 2 metres down, coming to the surface at night or during heavy rain.

Structure: see Fig. 15.24. Up to 30 cm long, the body is divided into distinct segments. Essentially the body consists of two tubes, one inside the other; the body wall on the outside, the alimentary canal on the inside, and the two separated by a fluid-filled cavity which acts as a

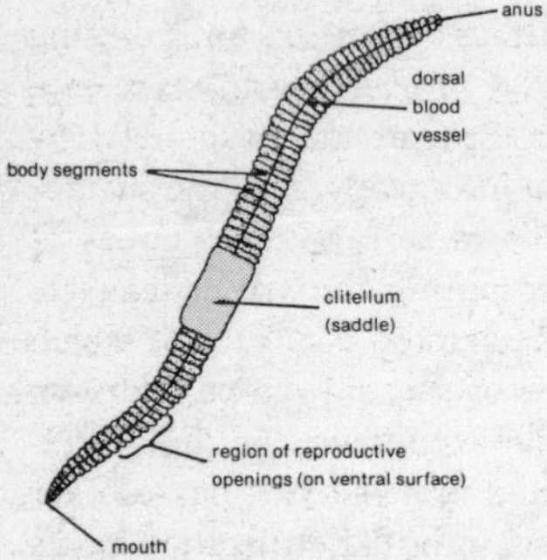

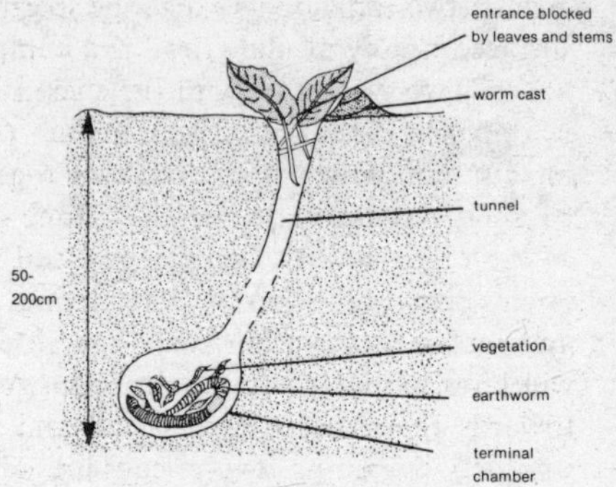

Fig. 15.24. (*a*) *Dorsal view of earthworm* (*b*) *Earthworm in its burrow*

hydrostatic skeleton (see page 190). There is a true gut with mouth and anus. Bristles called **chaetae** can be extended from the ventral surface and are used in locomotion.

Nutrition: holozoic. Vegetation and soil particles are ingested through the mouth and pass along the gut under muscle action. Digestive enzymes are secreted by the gut lining and the soluble products of digestion are absorbed into blood capillaries for transport to all cells in the body. Undigested food and soil particles pass out via the anus to form the familiar worm cast.

Respiration: aerobic.

Gas exchange: as in *Amoeba*, gas exchange in the earthworm takes place by simple diffusion across the body surface. In the earthworm, however, a circulatory system is necessary to deliver the oxygen to all internal tissues. The skin of the earthworm is a specialized respiratory surface (see page 102).

Movement: changes in body shape are brought about by the action of antagonistic circular and longitudinal muscles acting on the fluid skeleton. When circular muscles contract, the body becomes thin and long; when longitudinal muscles contract, the body becomes short and fat. During forward movement through the burrow or on the soil surface, alternate contractions and relaxations start at the head and pass back along the body. The head is extended, then fattens, with its chaetae anchored in the soil, and then the rest of the body is drawn up towards it. If the soil is tightly compacted, soil is steadily ingested as the worm moves forwards. The worm produces mucus (slime) to lubricate the sides of the burrow.

Reproduction: the earthworm is hermaphrodite, but cross-fertilization occurs (two individuals exchange sperms). Mating takes place at the soil surface, usually at night, and is a complex process which takes about 3 hours. Two worms lie with their ventral surfaces together, but pointing in opposite directions. The clitellum (saddle) of each worm secretes a mucus tube which binds the two together. Sperms are exchanged by passing from the male opening of one worm into grooves on the surface of the other worm. The grooves lead to sperm storage sacs. Tubes of mucus prevent the sperms from mixing. After mating, the worms separate and return to their burrows. The clitellum secretes a substance which will later harden to form a cocoon. A tube of this substance is drawn towards the anterior (head) end and as it passes over the female reproductive opening several eggs are deposited in it. Next it passes over the sperm storage sac and sperms are passed into it. Fertilization takes place inside the cocoon, and as it slips off the head end, the ends seal. After about 12 weeks, *one* young worm emerges from the cocoon.

Importance of earthworms

By their actions earthworms improve the fertility of soil:

1. Their burrows aerate, drain and break up the soil.
2. Their casts, which contain fine soil particles, help produce a good tilth (seed bed). Their action also mixes soil from lower layers with that on the surface, thus increasing the depth of topsoil.
3. The worm's waste products add manure to the soil and encourage the activity of soil bacteria.
4. Pieces of vegetation, e.g. leaves, dragged down into burrows help to mix humus with the soil.

15.18. *Arthropods*

Phylum: Arthropoda

Features of the group:

1. Body covered by **exoskeleton**
2. Body divided into segments
3. Jointed limbs

The arthropods are divided into four classes:

1. **Myriapods:** body made up of many similar segments each bearing appendages (e.g. centipedes and millipedes)
2. **Crustacea:** two pairs of antennae. Most live in water (e.g. crab, lobster, water flea); some live on land (e.g. woodlouse)

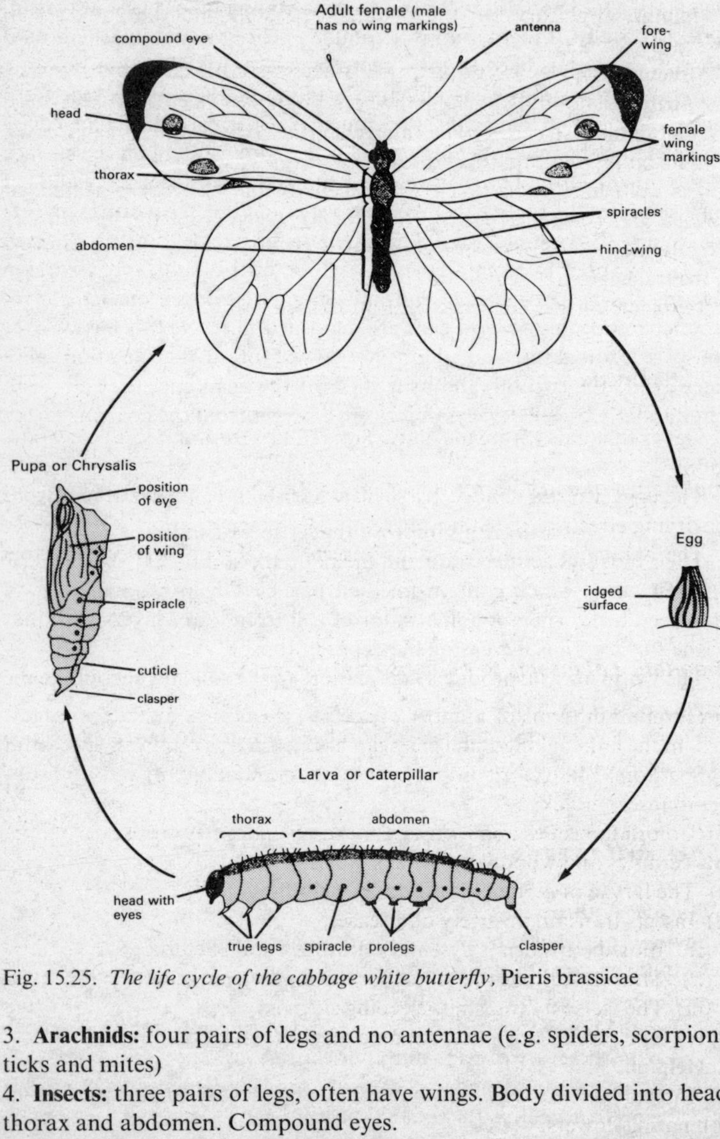

Fig. 15.25. *The life cycle of the cabbage white butterfly,* Pieris brassicae

3. **Arachnids:** four pairs of legs and no antennae (e.g. spiders, scorpions, ticks and mites)
4. **Insects:** three pairs of legs, often have wings. Body divided into head, thorax and abdomen. Compound eyes.

The Cabbage White Butterfly, **Pieris brassicae** – an example of an insect

Habitat: migratory but often seen in gardens and rural regions in warm, dry weather.

Structure: see Fig. 15.25

Nutrition: The adults suck up the nectar of flowers through a **proboscis**, a long tube-like structure. The caterpillar (**larva**) feeds on cabbage leaves with its biting mouthparts.

Gas exchange: air is taken to the cells of the body by a system of trachea and tracheoles (see section 6.4). Oxygen from the air diffuses directly into the cells, carbon dioxide diffuses from the cells into the air in the tracheoles.

Reproduction and life cycle: see Fig. 15.25.

Adaptations to living on land

1. The cuticle is impermeable to water.
2. Gas exchange is within the body. This reduces the water loss by evaporation.
3. The eggs are surrounded by shells which help to protect them from the drying effects of the atmosphere.
4. The ability to fly allows many insects to migrate between suitable climates.

Importance of insects to humans

1. **Harmful:** in terms of number of species, insects are by far the largest class in the animal kingdom. They can be regarded as our most successful competitors. Many agricultural crops and manufactured goods are subject to insect attack:

(a) Colorado beetles and locusts, for example, destroy crops.
(b) Termites and woodworm attack wood.
(c) The larvae of certain moths destroy clothing.
(d) Insects transmit a variety of diseases:
  (i) Blowflies transmit dysentery protozoa and bacteria.
  (ii) The mosquito is the vector for the malaria parasite.
  (iii) The tsetse fly transmits sleeping sickness.
  (iv) The rat flea transmits bubonic plague (black death).

2. **Helpful:**

(a) Insects such as bees and butterflies play a very important role in pollinating flowers.
(b) Insects which are natural predators of pests can be used for biological control, e.g. ladybirds feeding on aphids.

(c) Bees provide wax and honey.
(d) Silkworms (larvae of the silkworm moth) provide silk.

15.19. *Molluscs*

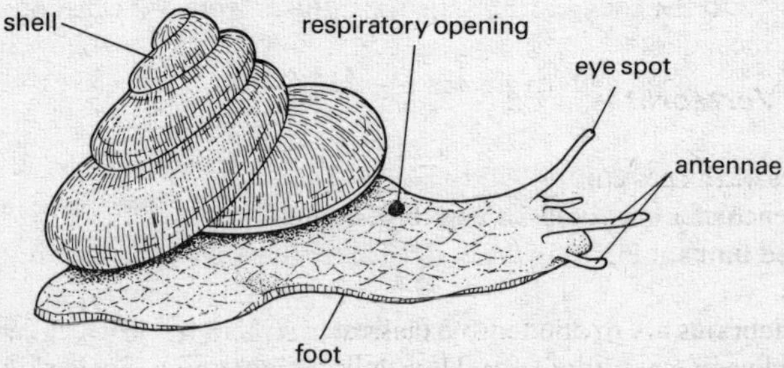

Fig. 15.26. *External structure of a land snail*

Phylum: Mollusca. This phylum includes the snails, slugs, winkles, whelks, mussels, oysters, octopuses and squids. We shall consider a land snail as a representative of the group.

Land snails

Structure: see Fig. 15.26

The head bears 2 pairs of tentacles. The upper pair of tentacles possess eyes. The large muscular foot provides movement. Waves of muscular contraction pass along it, moving the snail along. The foot is lubricated by mucus produced by slime glands in its surface.

Nutrition: snails feed on a wide range of plant material. They have hard, rasping mouthparts which are used to scrape food from the surface or to cut through vegetation.

Gas exchange: air passes in and out through the respirator pore and gas exchange takes place within the shell.

Reproduction: snails are hermaphrodite, that is each snail possesses both male and female sex organs. When two snails mate they fertilize each other at the same time. Eggs are laid in clutches in holes in the soil or in cracks and crevices. The young hatch as miniature versions of the adult.

Adaptations to living on land

The greatest problem for the snail is to conserve water. This is achieved by:

1. Withdrawing the body into the shell in dry conditions.
2. Having gas exchange surfaces within the shell.
3. Having the foot covered with mucus.
4. Choosing damp, shady habitats.
5. Being most active at night to avoid the heat of the day.

15.20. *Vertebrates*

Features of the group:

1. Presence of a backbone or vertebral column.
2. Paired limbs.
3. Jaws.

The vertebrates are divided into 5 classes:

1. **Fish:** live in water (aquatic). Have gills for breathing. Scaly skin. Fins for movement. Skeleton may be made of cartilage (sharks) or bone (most fish, e.g. herring).
2. **Amphibia:** live on land but return to water to breed. Moist skin (e.g. frogs, toads, newts, salamanders).
3. **Reptiles:** dry, scaly, waterproof skin. Eggs have a leathery shell and are laid on land (e.g. snakes, lizards, turtles, crocodiles).
4. **Birds:** covered with feathers. Forelimbs are modified to form wings. No teeth, mouthparts modified to form beak. Endothermic (warm blooded).
5. **Mammals:** body covered in hair. Young are fed on milk from mammary glands. Ears are external. Sweat glands are found in the skin. The body is divided by the diaphragm into the thorax and abdomen.

Some syllabuses required you to know about certain vertebrates in more detail. These are described in the following pages.

15.21. *A Fish: the Herring,* Clupea harengus

Phylum: Chordata (chordates).
Sub-phylum: Vertebrata (vertebrates).
Class: Pisces (fish).

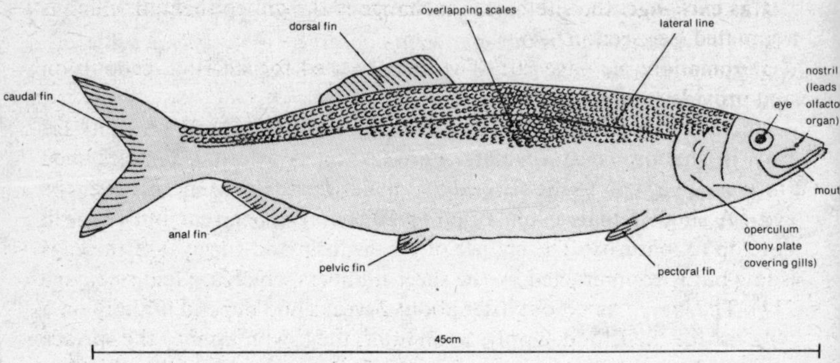

Fig. 15.27. *The herring,* Clupea harengus

There are two groups of fish: bony fish (those with a bony skeleton) and cartilaginous fish (those with a skeleton of cartilage).

Features of bony fish (most are adaptations for an aquatic life):

1. The body is usually streamlined and has fins.
2. Overlapping bony scales covered in mucus provide protection and streamlining.
3. The organs of gas exchange are gills.
4. In most, movement is brought about by waves of contraction that pass along muscle blocks (myotomes) on either side of the bony vertebral column (see section 10.5).
5. In most, a lateral line is used for detecting vibrations on the water.
6. Fish are ectothermic (have a variable body temperature (see page 159)).
7. Sexes are separate and fertilization is usually external.

Habitat: marine, lives in waters around the British Isles.

Structure: see Fig. 15.27. Examine the external features as adaptations for aquatic life.

Nutrition: holozoic. They are carnivorous, feeding mainly on copepods (small crustacean arthropods) which they filter out of the water using gill rakers.

Excretion and osmoregulation: water leaves the fish by osmosis (page 40) and is replaced by drinking sea water and then actively expelling the salts it contains (see page 149). Soluble nitrogenous wastes are excreted in the urine produced by the kidneys.

Respiration: aerobic.

Gas exchange: the site of gas exchange is the gill epithelium which is ventilated (see section 6.4).

Locomotion: see page 201. The fins are used for steering, propulsion and provide stability.

Reproduction and life history: spawning occurs in autumn on stony sea beds in shallow coastal waters. Fertilization is external; females shed ribbons of eggs on to the sea bed and males then release sperms over the eggs. A single female can lay up to 20,000 eggs per year but few will develop to maturity. The chance of fertilization and survival of the eggs is low but is counteracted by the sheer numbers which are laid (see page 211). The larvae hatch out after about 2 weeks and depend initially on a yolk sac for their food supply. Soon after, they swim up into the surface waters where they feed and remain for 3–6 months while they develop into adults.

15.22. *An Amphibian: the Common Frog,* **Rana temporaria**

Phylum: Chordata (chordates).

Sub-phylum: Vertebrata (vertebrates).

Class: Amphibia (amphibians).

Features of the group:

1. Ectothermic vertebrates (see page 159) with paired limbs.
2. Adult terrestrial, larva aquatic.
3. Adult has lungs for gas exchange, larva has gills.
4. Skin of adult usually thin, moist and mucus-covered.
5. Adults return to water to breed and eggs are laid and fertilized in the water.
6. Gradual **metamorphosis** (see page 244) from larva to adult occurs.

Habitat: moist, shady places or in ponds and streams.

Structure: see Fig. 15.28.

Nutrition: holozoic. Adult feeds mainly on slugs, snails and insects which are caught by the rapid ejection of the frog's long sticky tongue. In the young larva (tadpole) a yolk sac provides the food supply till the mouth develops. Once the mouth is formed, the tadpole feeds on the covering of algae growing on the surface of stones and large plants.

Respiration: aerobic.

Gas exchange: newly hatched tadpoles have external gills for gas exchange. 2 to 3 weeks after hatching the gills become internalized and water is pumped over them between mouth and spiracle (see Fig. 15.28).

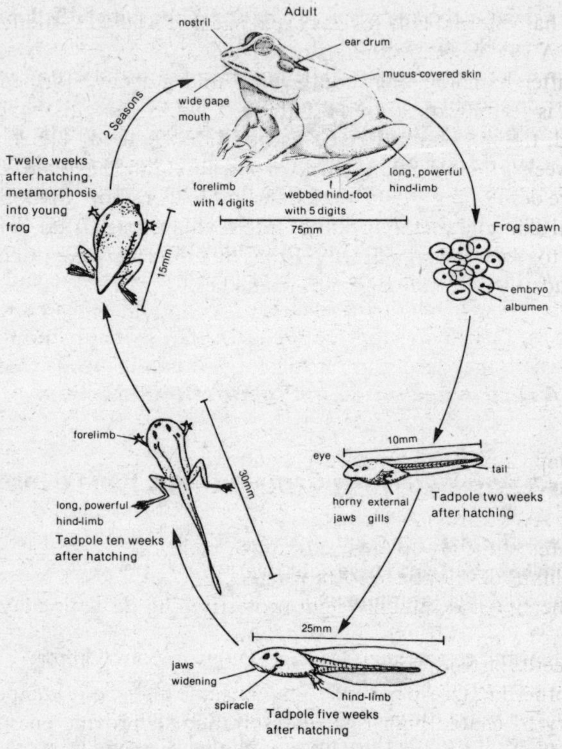

15.28. *Life cycle of the common frog*

In the adult, the mouth, lungs and skin are used for gas exchange (see section 6.4).

Locomotion: the hind-limbs of the adult are powerful for jumping on land and for swimming, assisted by the webbed feet. The short forelimbs act as shock absorbers on landing and are steering aids when swimming.

Reproduction and life history: frogs mate in early spring, the male on top of the female. Fertilization is external, the male releasing sperm over what may be 100 newly laid eggs. Coupling in this way increases the likelihood of fertilization.

The eggs swell soon after contact with water and the fertilized egg becomes surrounded by 'jelly'. The 'jelly' provides buoyancy and helps give protection against injury and predation. The embryo feeds on the jelly and hatches out about 2 weeks after fertilization. The newly hatched tadpole has external gills for gas exchange and feeds on an internal yolk supply. A sucker on its underside enables it to attach to weeds. By 2 weeks after hatching, the mouth, anus and gut have formed and the tadpole is feeding on algae. By about 5 weeks the gills have become enclosed, the hind-limbs appear as buds and the body has lengthened. By 10 weeks, the tadpole is carnivorous, forelimbs have appeared, and lungs are developing. At 12 weeks, the tadpole becomes immobile, sheds its larval skin and metamorphoses into a young frog. It takes two more seasons to reach sexual maturity. Frogs hibernate in winter, under stones or in mud at the bottom of ponds or ditches.

15.23. *A Bird: the Domestic Pigeon,* Columba livia

Phylum: Chordata (chordates).
Sub-phylum: Verbrata (vertebrates).
Class: Aves (birds).
Features of the group (adaptations for flight):

1. Forelimbs developed to form wings.
2. Feathers provide a lightweight protective and insulating layer on the body.
3. No teeth (instead, a beak).
4. Endothermic (see page 159) – a constant high body temperature is necessary to enable high rate of respiration to provide energy during flight.
5. Highly developed nervous system enables remarkable muscular co-ordination.

Habitat: common in towns and cities.

Structure: see Fig. 15.29.

Nutrition: holozoic. Feeds on seeds, young green shoots and occasionally insects and snails.

Respiration: aerobic.

Gas exchange: ventilated lungs and air sacs are used. An efficient system is required because of the high rate of respiration during flight.

Locomotion: flapping flight.

Reproduction: fertilization is internal, the male mounting the female after courtship display. Sperms from the male are deposited in the oviduct of the female. Yolk, albumen, membranes and a hard shell develop around the fertilized egg. The resulting hard-shelled ‘egg’ (Fig. 15.30) continues down the oviduct and is laid in a nest constructed by both

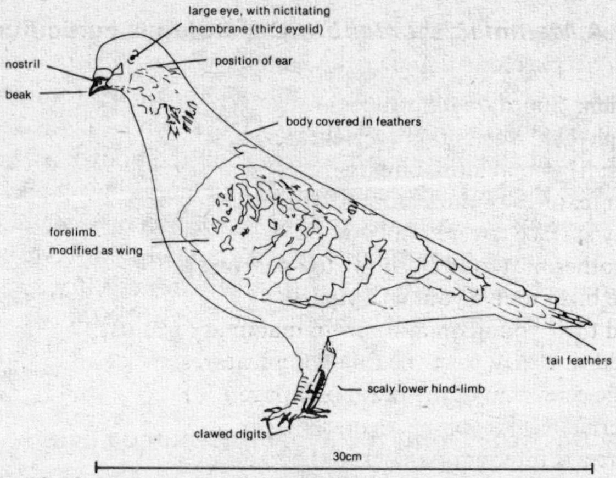

Fig. 15.29. *The domestic pigeon*, Columbia livia

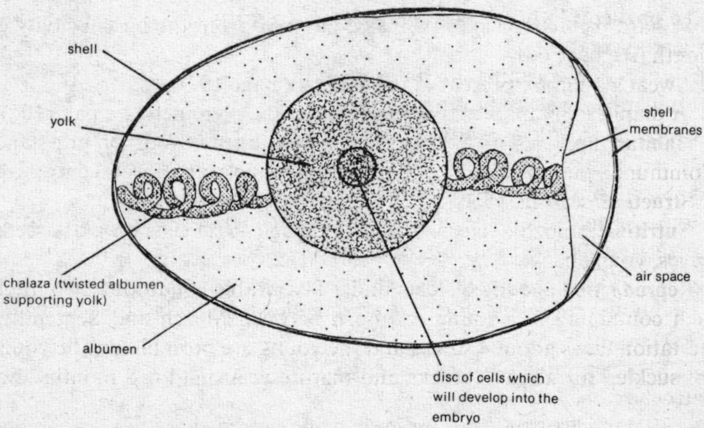

Fig. 15.30. *A newly laid fertilized bird's egg*

parents. The developing embryo depends on yolk for its carbohydrate and fat supplies; albumen provides protein and water. Wastes are either deposited in a solid form, e.g. uric acid, or diffuse out through the egg shell, e.g. carbon dioxide. Oxygen is obtained by diffusion through the egg shell.

The egg is incubated (kept warm) by both parents in turn. The two eggs in a clutch hatch after about 18 days. The young feed initially on 'pigeon's milk', a white secretion regurgitated from the mother's crop. The young fly at 35–37 days and leave the nest soon after.

15.24. *A Mammal: the Rabbit,* Oryctolagus cuniculus

Phylum: Chordata (chordates).
Sub-phylum: Vertebrata (vertebrates).
Class: Mammalia (mammals).
Main features of the group:

1. Body surface covered in hair.
2. Endothermic (constant body temperature).
3. Give birth to live young (viviparous).
4. Feed their young on milk, from mammary glands.

In addition they have the following features:

1. Large cerebrum of the brain (see page 170).
2. External ear, the pinna, is present.
3. There are three ear ossicles in the middle ear.
4. Heterodont dentition – different types of teeth are present (see page 83).
5. A hard palate separates the nasal passages from the buccal cavity or mouth (see page 66).
6. Sweat glands are present in the skin (see page 157).
7. A diaphragm separates the thorax from the abdomen (see page 105).

Habitat: burrows in light, sandy soil in open woods or grassland. Communal burrows consist of connecting tunnels with several entrances.

Structure: see Fig. 15.31.

Nutrition: holozoic. Herbivorous (see page 47). Feeds on grass seeds, leaves, young bark.

Reproduction and life history: under favourable conditions, 3–6 litters each containing 3–7 young are born between March and September. Gestation takes about 4 weeks and the young are born blind. The young are suckled for about 4 weeks and mature at around 6–8 months. For

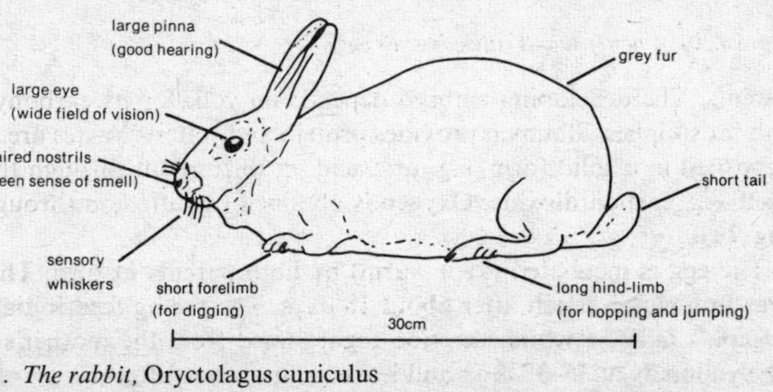

15.31. *The rabbit,* Oryctolagus cuniculus

most individuals, because of predation, the life span is around one year.

Their very high reproductive rate, resulting from short gestation, many litters and early sexual maturity sometimes makes rabbits a pest on agricultural land. The virus disease myxomatosis was introduced to control them, but some rabbits are now immune to this disease.

16. In the Exam

If you followed the advice and instructions in Chapter 2 during your revision, you should now be familiar with the different types of question set by your board and should know how to tackle them. In any case, read over Chapter 2 once more.

Get plenty of sleep the night before your exam. Do *not* stay up all night revising. Get up in good time for the exam, allowing extra time for travel in case of unforeseen delays (traffic jams, train cancellations, a bicycle puncture, etc.). Turn up to the exam in good time and come fully equipped with pen, pencils, pencil sharpener, eraser and ruler.

Now is the time to put all your effort to good use. Remember, what counts is what you put down on the exam paper. This is your only contact with the examiner. Marks can only be given for what you write on the paper – not for what remains in your head.

1. Read carefully the instructions at the head of the exam paper and follow them absolutely.
2. Work out a time allocation for the questions according to the marks they carry. Throughout the exam, check on the time and stick to your time allocation – do not overrun.
3. Answer all compulsory questions.
4. If the paper is divided into sections make sure you answer the correct number of questions from each section.
5. Always answer the correct *number* of questions. If you are required to do four long-answer questions and you only answer three, it is almost impossible to make up the marks for the missing fourth question. The first 50 per cent of marks are the easiest to gain on any question. To get 60 marks out of 100, it is easier to gain 15 out of 25 on each of four questions than to gain 20 out of 25 on each of three.
6. If given a choice of longer-answer questions, narrow down the best questions to answer. You can do this by estimating the number of marks you will get for a question *before* you attempt to answer it.
7. *Understand* what the question asks *before* you attempt to answer it. Do *not* twist the question into an unintended meaning. *No* marks are awarded for irrelevant information and the answer should include *only*

points demanded by the question. Do not write down *all* you know on a particular topic unless it is asked for.

8. Examiners prefer concise answers written in good English. Set out work neatly – the examiner is only human. Draw a single straight line across any work which you do not wish to be marked (an essay plan for example).

9. Plan your longer answers before writing them (see page 9).

10. Use large labelled diagrams if they make your answer clearer and save time and words.

11. If you are running out of time in a question, complete your answer in note form, putting in all the key points. You will gain at least some marks.

12. If any time is left at the end, read through your answers and correct any mistakes.

Look on the exam as a culmination of your efforts, a pinnacle you have reached rather than a dreadful hurdle you must haul yourself across.

PENGUIN REFERENCE BOOKS

The Penguin English Dictionary

Over 1,000 pages long and with over 68,000 definitions, this cheap, compact and totally up-to-date book is ideal for today's needs. It includes many technical and colloquial terms, guides to pronunciation and common abbreviations.

The Penguin Reference Dictionary

The ideal comprehensive guide to written and spoken English the world over, with detailed etymologies and a wide selection of colloquial and idiomatic usage. There are over 100,000 entries and thousands of examples of how words are actually used – all clear, precise and up-to-date.

The Penguin English Thesaurus

This unique volume will increase anyone's command of the English language and build up your word power. Fully cross-referenced, it includes synonyms of every kind (formal or colloquial, idiomatic and figurative) for almost 900 headings. It is a must for writers and utterly fascinating for any English speaker.

The Penguin Dictionary of Quotations

A treasure-trove of over 12,000 new gems and old favourites, from Aesop and Matthew Arnold to Xenophon and Zola.

The Penguin Wordmaster Dictionary Manser and Turton

This dictionary puts the pleasure back into word-seeking. Every time you look at a page you get a bonus – a panel telling you everything about a particular word or expression. It is, therefore, a dictionary to be read as well as used for its concise and up-to-date definitions.